LENSES ON LEARNING
MODULE 2

Teacher Learning for Mathematics Instruction

Facilitator Book

Lenses on Learning Project Staff
Catherine Miles Grant, Barbara Scott Nelson, Ellen Davidson,
Annette Sassi, Amy Shulman Weinberg, Jessica Bleiman

Center for the Development of Teaching
Education Development Center
Newton, Massachusetts

DALE SEYMOUR PUBLICATIONS
Pearson Learning Group

 **National Science Foundation**

This work was supported by the National Science Foundation under Grant No. ESI-9731242 and by The Pew Charitable Trusts. Any opinions, findings, conclusions, or recommendations expressed here are those of the authors and do not necessarily reflect the views of these organizations.

Special thanks are due to our editor, Beverly Cory. Her understanding of the ideas in these materials and their pedagogical stance, coupled with her commitment to clarity, made the materials much more understandable and usable.

Art and Design: Bernadette Hruby, Senja Lauderdale, Jim O'Shea

Editorial: Doris Hirschhorn

Manufacturing: Mark Cirillo, Sonia Pap

Marketing: Douglas Falk

Production: Karen Edmonds, Alia Lesser

Publishing Operations: Carolyn Coyle, Tom Daning, Richetta Lobban

Copyright © 2003 by the Education Development Center, Inc. Published by Dale Seymour Publications®, an imprint of Pearson Learning Group, 299 Jefferson Road, Parsippany, NJ 07054. All rights reserved. No part of this book may be reproduced or transmitted in any form or by any means, electronic or mechanical, including photocopying, recording, or by any information storage and retrieval system, without permission in writing from the publisher. Blackline masters excepted. For information regarding permission(s), write to Rights and Permissions Department.

ISBN 0-7690-3027-0

Printed in the United States of America

1 2 3 4 5 6 7 8 9 10 07 06 05 04 03 02

1-800-321-3106
www.pearsonlearning.com

The *Lenses on Learning* Project

Project Staff, Education Development Center

Barbara Scott Nelson (PI)

Jessica Bleiman

Ellen Davidson

Catherine Miles Grant

Annette Sassi

Amy Shulman Weinberg

Sheila Flood

Pilot-Test Facilitators

Jeffrey Benson
Principal and Director of Educational Services
Germaine Lawrence School
Arlington, Massachusetts

Valerie Gumes
Principal, Blue Hills Early Education Center
Boston, Massachusetts

Joanne Gurry
Assistant Superintendent, Arlington Public Schools
Arlington, Massachusetts

Eric Johnson
Principal, Dever School
Boston, Massachusetts

Joseph Petner
Principal, Haggerty School
Cambridge, Massachusetts

Emily Shamieh
Principal, Winthrop School
Boston, Massachusetts

Debra Shein-Gerson
Elementary Mathematics Curriculum Coordinator
Brookline, Massachusetts

Casel Walker
Principal, Manning School
Boston, Massachusetts

These pilot tests took place in eastern Massachusetts under the auspices of the Boston Public Schools, Education Collaborative of Greater Boston (EDCO); Lesley College; and the Merrimack Education Center.

Field-Test Sites

Albuquerque Public Schools
Albuquerque, New Mexico

Clark County School District
Las Vegas, Nevada

Durham Public Schools
Durham, North Carolina

Greenville County Public Schools
Greenville, South Carolina

Mt. Holyoke College
South Hadley, Massachusetts

Madison Elementary School District #38
Phoenix, Arizona

SUNY-Cortland
Cortland, New York

University of Washington
Seattle, Washington

Project Evaluators, Education Development Center

Barbara Miller

Michael Foster

Sarah Gray

Project Advisors

Diane Briars
Pittsburgh Public Schools

Nancy Dickerson
Boston Public Schools

Judy Mumme
Mathematics Renaissance K–12

Joseph Murphy
The Ohio State University

Mildred Collins Pierce
Harvard Graduate School of Education

Susan Jo Russell
TERC, Cambridge, Massachusetts

James Spillane
Northwestern University

Virginia Stimpson
University of Washington

Philip Wagreich and Kathy Kelso
University of Illinois at Chicago

CONTENTS

About This Course 1
Facilitator's Notes for the Course 4
How to Use These Materials 6
Activity Guidelines for Facilitators 9
Introduction to *Module 2* 22

SESSION 1 Changing More Than One's Socks 29
- OPENING Welcome 32
- ACTIVITY 1 Dimensions of Change 34
- ACTIVITY 2 What Are Teachers Grappling With? 37
- CLOSING Bridging to Practice and Homework 41

SESSION 2 What Do Teachers Need to Learn? 47
- OPENING Introducing the Session 50
- ACTIVITY 1 One Professional Development Experience 51
- ACTIVITY 2 Teachers Learning the Mathematics They Teach 54
- ACTIVITY 3 Understanding Children's Mathematical Thinking ... 62
- CLOSING Bridging to Practice and Homework 67

SESSION 3 What Makes for Meaningful Professional Development? ... 73
- OPENING Getting Started 76
- ACTIVITY 1 Facilitating Discourse 77
- ACTIVITY 2 The Teachers in Our Schools 83
- ACTIVITY 3 Learning Mathematics Together 87
- CLOSING Bridging to Practice and Homework 93

SESSION 4 Critical Colleagueship 99
- OPENING Introducing the Session 102
- ACTIVITY 1 Exploring Critical Colleagueship 103
- ACTIVITY 2 Alternative Images of Professional Development .. 107
- CLOSING Bridging to Practice and Homework 112

SESSION 5 Providing Professional Development 115
- OPENING Bringing It All Together 118
- ACTIVITY 1 Professional Development in Our Schools 120
- ACTIVITY 2 Vignettes of Individuals Moving in the Stream ... 122
- ACTIVITY 3 A System on the Move: Project IMPACT 125
- CLOSING Bridging to Practice 129

Resource List 131

About This Course

The move to standards-based mathematics instruction has been underway in this country, and internationally, for more than a decade. This effort to improve mathematics instruction began with the 1989 publication of the *Curriculum and Evaluation Standards* of the National Council of Teachers of Mathematics, commonly termed the NCTM *Standards*, and has continued with the development of new standards and frameworks in most states and many school districts. More recently, NCTM has published *Principles and Standards for School Mathematics* (2000), a revision of the 1989 *Standards* based on a decade of experience in schools around the country. During this same period, many new mathematics curricula have been developed; these new programs give teachers the tools to teach mathematics in a way that encourages rigorous student thinking. There is a growing appreciation of the unique ways in which children develop mathematical understanding and an effort to tap into the range of strengths they bring from the cultures in which they live. New assessment instruments are being developed, too, which evaluate not just students' mastery of important facts and mathematical procedures, but also the way they *think* mathematically.

Standards-based instruction is based on a set of understandings about the nature of mathematical knowledge, children's learning, and teaching. These understandings have developed in the years since most current school and district administrators were educated. A large proportion of today's administrators were educated at a time when mathematics was viewed as an assemblage of facts and procedures, learning was considered the process of absorbing new information and practicing new skills, and teaching was seen as transmitting accumulated knowledge and giving students plenty of practice with new skills. However, we now view mathematics as a subject that can be reasoned out; learning as the process of reasoning things out, in interaction with the material and social world; and teaching as the process of facilitating students' mathematical reasoning.

Many administrators have found that acting on their older ideas about mathematics education, even with good intentions, often does not have the intended effect. They are eager for the opportunity to explore the new ideas in a relaxed setting, to examine them in concrete ways that are meaningful to them, and to discuss with colleagues the implications of these ideas for their own leadership roles. The course *Lenses on Learning* offers administrators that opportunity.

Content of the course

Lenses on Learning is a set of seminars in three modules, designed to help administrators, as instructional leaders in their schools and districts, think through the ideas that underlie standards-based reform in elementary mathematics and relate those ideas to their own work. While there are many such professional development opportunities for *teachers,* administrators rarely have the same chance to dig deeply into the new ideas, puzzling out for themselves what the changes are all about and thinking about the implications for their own administrative responsibilities, such as supervising teachers, communicating with parents, and navigating varied approaches to student assessment. Through *Lenses on Learning,* school- and district-level administrators consider the following topics:

- **The nature of mathematical understanding** Administrators engage in mathematical problem solving in important topics in the elementary mathematics curriculum, presented at an adult level. Through this work, they gain an appreciation for the nature of mathematical understanding that goes beyond algorithmic manipulation.

- **The development of children's mathematical understanding** Administrators read excerpts from research on the development of children's mathematical thinking, watch and discuss videotapes of clinical interviews with children, and examine children's written work in order to understand how children's mathematical thinking develops.

- **Discourse-based mathematics instruction** Administrators read descriptions of discourse-based elementary mathematics classrooms and watch and discuss videotapes of teachers in action. In this way, they work to develop an eye for standards-based elementary mathematics classrooms. They consider how the learning of mathematical concepts and the learning of mathematical facts and algorithms can happen simultaneously through problem solving. They also observe and discuss the teachers' instructional moves and consider new approaches to classroom observation and teacher supervision.

- **Professional development for teachers** Through readings, excerpts from teachers' journals, and videotapes of mathematics classrooms, administrators consider what teachers need to know, and what they need to know how to do, in order to facilitate the development of children's mathematical thinking. They are introduced to several approaches to professional development that are designed to address those needs.

Depending on their particular roles as administrators and instructional leaders, participants may also consider topics of special interest to them, such as the use of instructional technology or the relationship between computation and conceptual understanding.

Thinking in order to learn

This course does not cover the foregoing topics one by one. Rather, administrators explore these new ideas by actively working on intellectually interesting and practical, relevant issues. They are, in effect, thinking in order to learn. In each of three modules, administrators consider issues embedded in standards-based mathematics instruction as they emerge in particular aspects of their work. The "big ideas" participants consider in the course, however, are not related singly to particular administrative functions; most ideas are examined in the context of several functional areas. Encountering these ideas several times, in slightly different contexts, enriches the thinking that administrators do. The modules designed to expose these ideas are as follows:

- **MODULE 1 Instructional Leadership in Mathematics** What exactly are the new understandings about the nature of mathematics learning and teaching that underlie standards-based instruction? Participants explore fundamental issues and consider the implications for school leadership.

- **MODULE 2 Teacher Learning for Mathematics Instruction** What do teachers need to learn, and how can they get there? Participants focus on the issues teachers grapple with in learning to teach according to the new standards, and they develop a sense of the criteria for effective professional development across the school community.

- **MODULE 3 Observing Today's Mathematics Classroom** How can we tell what is going on in standards-based mathematics classrooms, and how can we tell when instruction is working? Participants consider what to look for when conducting classroom observations and what to discuss when meeting with the teacher about those observations.

Although it is not necessary to use the three modules in sequence, the first module provides a solid introduction to the new ideas to be considered throughout the course. During this first seminar, administrators have the opportunity to sort out their initial thoughts and reactions as they begin to think about the orientation they want to bring to their instructional leadership for changing mathematics education in their schools and districts. *Module 2* and *Module 3* treat these ideas in the context of specific functional aspects of administrators' work, in practical areas for which they have responsibility and in which they make judgments and take action on a regular basis. While the modules focus on different administrative functions, the ideas about mathematics learning and teaching are cumulative, building from one module to another. Participation in seminars covering either two or three modules can substantially deepen administrators' understanding of the ideas that undergird standards-based mathematics education.

Facilitator's Notes for the Course

As administrators think through the content of these modules, your job, as facilitator, is to attend to the ideas that participants express; to understand them as representations of each administrator's current view of learning, teaching, or mathematics; and to guide the discussion so that there is time and space for the respectful consideration of the range of ideas that may be present in the group. You are also there to offer encouragement to people who may be struggling with ideas that are new to them. Facilitator's notes, which appear in the introduction to each module as well as with each class session, can prepare you for the ideas administrators might express and help you shape your own contributions to foster the growth of administrators' understanding and perceptions.

You will find three themes woven throughout these facilitator's notes: (1) how to help administrators who are considering new ideas about mathematics education; (2) how to establish and maintain a reflective community in your seminar; and (3) how to help participants connect the new ideas to their administrative practice.

How to help administrators who are considering new ideas about mathematics education

By and large, administrators are most familiar with the traditional form of mathematics education that is currently under revision. Although they may have read a great deal in recent years about the NCTM *Standards*, about state and district mathematics frameworks, and about research on both children's learning of mathematics and diversity in schools, many of their deeply held ideas about mathematics learning and teaching come from their own experience as students in school, from what they were taught when they were in college and graduate school, and from what they experienced as teachers of elementary school mathematics themselves.

Most of today's administrators were in elementary and secondary school during the period when instruction was informed by a behaviorist perspective on learning. Ideas from the cognitive perspective that inform the current standards-based principles of mathematics instruction emerged only toward the end of their college or graduate school years, or even after they left. As a result, many administrators may view mathematical instruction as the learning of facts and procedures, and evidence of learning as the correct execution of these procedures or the application of them in word problems, rather than seeing mathematics learning as the gradual development of mathematical concepts, ways of thinking, and facts. Some administrators may have an amalgam of ideas from the behaviorist and cognitive perspectives. When administrators do mathematics together in this course, or when they discuss how children learn mathematics, their own particular amalgam of ideas is likely to come out. The facilitator's notes identify the ideas about mathematics learning and teaching that are highlighted in each particular session, provide examples of how administrators might talk about the ideas in that session, and give suggestions about how to direct classroom discussion to help administrators rethink important ideas.

How to establish and maintain a reflective community in your seminar

That classrooms should function as reflective communities for students and teachers is a hallmark of standards-based mathematics instruction. In standards and frameworks that call for communication and discourse, and in new mathematics curricula that advise teachers to ask students to explain their thinking, teachers are now enjoined to have students discuss their ideas, listen to each other respectfully, and think critically about the ideas expressed.

For the learning involved in this course, administrators, too, need to be part of a group that respects their ideas and supports their efforts to learn. Whether the group is discussing mathematics problems, a videotape of a mathematics classroom, or an assigned reading, interest in and respect for each other's ideas should be established as group norms. You, as facilitator, play an important role in helping participants be sensitive to one another's cultures and backgrounds as they work and talk together throughout the course.

Every attempt should be made to focus discussion on the *ideas* embedded in each reading or videotape, rather than on solving the administrative problems they apparently present. If you conduct the course so that your group functions as an analytic and reflective community, administrators will be free to think hard about significant and difficult issues, trusting that others will not judge or ridicule their tentatively held ideas. In such a community, participants will work to support and investigate each other's thinking. Establishing this climate in the seminar room helps administrators learn and, simultaneously, gives them a firsthand sense of how standards-based mathematics classrooms work. The facilitator's notes offer ideas for building a reflective community through the activities in each session.

How to help participants connect the new ideas to their administrative practice

Those experienced in professional development for teachers report that it is not really accurate to say that teachers learn new ideas and concepts in workshops or summer programs and then apply these ideas in their practice. Rather, the new ideas they pick up are like lenses through which teachers look at their classrooms, see things in a new way, act differently than before, and then interpret what happens (Schifter and Fosnot 1993). Their very understanding of the meaning of each new concept gets worked out and refined in the enactment. That is, "thinking in order to learn" happens in the context of daily practice as well as in workshops or summer institutes.

This is true for administrators as well. Throughout this course, participants have many chances to consider and discuss how the new ideas they are encountering relate to their ongoing administrative practice. There are homework assignments that require observing in mathematics classrooms in their own schools or that require interviewing teachers or students. Furthermore, the discussion of assigned readings or videotapes of mathematics classrooms often brings up related issues in the administrators' own practice. Making these connections opens up for discussion the assumptions that underlie both traditional and standards-based practice and encourages administrators to take practical action in their schools and districts. Facilitator's notes will help you achieve the appropriate balance between thinking and doing in the context of each session. The Bridging to Practice reflective writing that is done at the end of every class session offers a structured way to support participants in linking the ideas they are learning to their administrative practice.

How to Use These Materials

The three *Lenses on Learning* modules offer valuable work for administrators on fundamental ideas about mathematics learning and teaching. The way these big ideas emerge is slightly different in each module because of the "angle of vision" of that module's applied topic. As a facilitator, you will be working with one or more of the following modules:

- **Module 1** Instructional Leadership in Mathematics
- **Module 2** Teacher Learning for Mathematics Instruction
- **Module 3** Observing Today's Mathematics Classroom

A module contains four or five class sessions, each expected to run about three hours. Ideally, a "core" experience for administrators would be some version of the first module plus either the second or third module, or both.

We recommend that the experience begin with two to four sessions from *Module 1*. Facilitators can make judgments about how much of this module is necessary, depending on the experience that participating administrators have already had. For example, if administrators in your group are from districts in which standards-based mathematics curricula have been in use for several years and there has been a rich program of professional development for teachers, such a group may be familiar with many of the ideas in *Module 1* and will need only one or two sessions to get a feel for what is expected of them in the *Lenses on Learning* course. If you are offering this course in a university environment and the students have already had courses that cover the content of *Module 1,* you may want to start with *Module 2* or *Module 3*. All subsequent modules presume the experience in the introductory module, including familiarity with being in a reflective community that explores ideas about mathematics learning and teaching.

Structure of each module

The materials for each *Lenses on Learning* module consist of two books and a videocassette. The facilitator's book contains the activity descriptions, related facilitator's notes, and handouts for that module. The readings are contained in a book that includes an introduction to the module and all the reading material required for class work and homework: articles and excerpts from books representing the educational research community, curriculum excerpts, video transcripts, and so forth. You will get, as a "facilitator's package," both the facilitator's book and the readings for each module you plan to teach, along with a videocassette to be shown in class during that module. Each participant in the seminar will need an individual copy of the readings; this book should be sent to them with the first homework assignment shortly before your seminar is scheduled to begin.

The facilitator's book is your guide for the course. Introductory material explains the *Lenses on Learning* course in general and then details the ideas covered in that particular module. One key section repeated in each module is Activity Guidelines for Facilitators (pp. 9–21); these pages explain your role as facilitator and discuss decisions you must make as you facilitate the kinds of activities that occur in *Lenses on Learning*: doing mathematics, viewing videotapes, examining students' and teachers' work, and holding small- and whole-group discussions.

After the introductory material, the remainder of the book is organized by class session. The opening three pages of each session offer an overview to help you prepare for the work you and the participants will be doing. This overview includes the following:

- Brief, introductory text that describes the session in general, provides the rationale for the work, and explains its niche in the larger context of the module
- A chart showing the overall structure of a session, with a synopsis of the activities, their purpose, and the suggested time allotment for each one
- The "big ideas" that administrators will be encountering in the activities of the session
- Materials you will need for the session, including handouts you will need to duplicate from this book
- Preparation for teaching the session: articles you need to read, math problems you should try doing yourself, video clips you need to preview, and so forth

The rest of the session alternates step-by-step descriptions of the activities with facilitator's notes that can help you work with your group. From two to four activities are suggested for each three-hour class session. The suggested time allotments for each activity are guidelines only; you may adjust these according to the nature and needs of your group. Within an activity, text headed *The Big Picture* briefly describes the intent of what participants are doing and underscores how the activities relate to the course as a whole; the purpose is to help you keep track of the big ideas that should be emerging through the participants' work. Each class session ends with a 15-minute Bridging to Practice, when administrators engage in reflective writing in their journals and homework is assigned.

The facilitator's notes offer important background information, highlight the ideas that readings or discussion questions are meant to raise, alert you to the issues that these activities might raise for administrators, and suggest ways in which you, as facilitator, might respond. These notes may include summaries of the video clips and the readings being used. Blackline masters for the handouts are found at the end of each session. These handouts contain discussion questions, notes for video reference, worksheets for some activities, and homework assignments.

The videocassette for each module contains all the clips to be viewed during the seminar. When you are asked to show a particular clip twice, it appears a second time on the videotape so that no rewinding is necessary. You should view the video clips several times before the session in which you will show them.

Establishing a schedule for your course

These materials have typically been used in yearlong courses, meeting one evening for three hours every three or four weeks. It works well to offer *Module 1* (four sessions) in the fall semester and either *Module 2* or *Module 3* (four or five sessions) in the spring semester. When *Lenses on Learning* is used in an in-service context, three-week intervals between class sessions are optimal. If classes meet every two weeks, the pace of the course can seem rushed for both facilitators and participants; and if classes meet monthly, it can be hard for both facilitators and participants to maintain focus through the intervening weeks of busy administrative life. Nevertheless, this course has been successfully taught on a weekly basis in a university context in which both facilitators and participants are familiar with and committed to a faster working pace. With some modifications, the *Lenses on Learning* modules could be offered as a summer institute or as a series of three-day institutes during the school year.

Using the materials for each class session

Set aside several hours to prepare for each class session. Read the facilitator's notes that accompany that session, and also reread the relevant parts of Activity Guidelines for Facilitators for tips on doing mathematics, guiding discussions, viewing videotapes, and looking at students' or teachers' work.

At the beginning of every class, be sure to tell participants the focus of the upcoming session and explain how this work will support the ideas that are unfolding throughout the module. It is helpful to post an agenda of the day's activities in a place that is clearly visible. In that way, participants can share in the responsibility of seeing that the sessions' goals are kept in view. As noted previously, the amount of time each activity might take is meant to serve as a guideline, not a blueprint, as you plan the class. You may make decisions to change the flow of activities, either in advance or as the session progresses, to bring out the big ideas and to meet the particular needs of your group.

ACTIVITY GUIDELINES FOR FACILITATORS

In order to establish a reflective and analytic community in your seminar, you need to build an atmosphere of trust and commitment among participants. This begins with setting certain norms for course participation. Participants should show each other respect by arriving on time, ready to work; they should have done the homework; and if they have to miss a session, they should let you know in advance so that you can convey their regrets to the whole group. All of these courtesies help to build a sense of community in which participants can count on one another's presence and attentiveness as they work through new ideas together. You might make these commitments clear to participants as they sign up for the course.

To ground participants' work with new ideas in real contexts, the course involves them in several different types of activities. They do mathematics together; engage in small- and whole-group discussions; watch and discuss videotapes that reveal children's mathematical thinking and the nature of standards-based classrooms; examine samples of children's mathematical work and excerpts from teachers' professional journals; discuss the challenges of their own administrative practice; and write in their journals. The paragraphs that follow offer guidelines for facilitating each type of activity. This text is repeated in the introductory section of each module to encourage you to read it anew before each seminar or simply refer to it as you prepare for individual sessions. Either way, your growth as a facilitator will lead you to find new ideas and make new connections even in text you have read before.

Doing mathematics together

In many class sessions, administrators have mathematics problems to work on in small groups. Why does a professional development course for administrators ask them to do mathematics problems? What do they gain from the experience? Why were these particular mathematics topics chosen?

Most administrators were educated in a very traditional form of mathematics instruction and, as teachers, likely taught mathematics in a traditional way. While they probably have read many of the NCTM and state frameworks documents and are familiar with such terms as *problem solving* and *mathematical discourse,* as well as the idea that students should "explain their thinking" in mathematics classes, many administrators may not have concrete experiences to link to those words and, therefore, may interpret them in a traditional way. Furthermore, they may never have experienced mathematics as a terrain of interesting ideas to think about, and they may never have seen themselves as mathematical thinkers.

In its traditional form, mathematics education emphasizes learning mathematical facts, such as the multiplication tables, and procedures or algorithms, such as borrowing or long division. Although these facts and procedures are important, focusing solely on them in mathematics instruction often masks the underlying ideas about numbers and their relationships that actually make the procedures work. Without understanding those underlying ideas, a person must rely on memorizing the facts and procedures. Yet, as we all know, memory fails. When this happens, a person who understands the underlying relationships between numbers can reconstruct adequate procedures for solving a problem.

While doing mathematics in this course, administrators are exploring numbers and their relationships. The topics they consider—including two-digit subtraction and fractions—are important elements of the elementary school mathematics curriculum, giving administrators the chance to reacquaint themselves with concepts that teachers in their schools are teaching and talking about. The activities are specifically designed to put administrators in touch with themselves as mathematical thinkers as they puzzle about complex and interesting ideas. The structure also invites administrators to listen to one another's mathematical ideas, articulate their own, reflect on the values embedded in their ideas, engage in collaborative problem solving, and experience what it is like to be part of a group of people working on mathematical ideas together. Thus, doing mathematics cooperatively gives administrators a contemporary experience with mathematics learning while also building a reflective community in your classroom.

These experiences are critical if administrators are to understand the goals of standards-based mathematics instruction, the kind of mathematical understanding that all children are expected to develop, and the way that teaching for understanding works. Note, however, that the mathematics problems in this course are not problems for children. Although they derive from topics in the elementary curriculum, they reflect an adult perspective. Administrators are expected to explore these mathematical problems and ideas to develop their own understanding, and should not view them as examples of the work children might do in classrooms.

Many administrators in your course may have disliked mathematics as students and may now be apprehensive about doing mathematics with their peers. Others may have always liked mathematics. Some, such as mathematics coordinators, have mathematics at the center of their professional attention. You will need to help such a mathematically heterogeneous group work together in a way that is productive for everyone. From the outset, you need to acknowledge this heterogeneity and point out the importance of working together on the mathematics. Emphasize the following five specific guidelines every time your group does mathematics:

- **You are not alone.** Remind participants that if some of them feel bad about math now, or if they felt bad in math class when they were growing up, it is not or was not their fault, and they are not alone. Mathematics has not been taught well for many decades; this is one of the reasons math education is changing.

- **Listen.** Whether doing mathematics together in small groups or discussing it in a whole-group configuration, participants should be attentive to everyone's contribution. Participants should give others in the group time to articulate fully what they want to say, and listen carefully so that they really understand what is said—well enough that they could either paraphrase the comments or ask a question to get further clarification. Every expressed opinion, idea, question, and answer is important and adds to everyone's learning.

- **Be supportive.** Both you, as facilitator, and the participants in your course should be supportive of one another so that people feel free to share ideas that they are not sure about, and even ideas that they suspect might be wrong. One way to show support is to offer suggestions constructively. Those who are already comfortable with mathematics can help those who are not by showing interest in their mathematical thinking. Often, even the most mathematically adept will discover that others have insights that are new to them.

- **Be able to disagree respectfully.** A climate of mutual respect will encourage people to disagree in reasonable ways.

- **Be courteous.** It is never acceptable to use sarcasm, to humiliate people, or to put down their ideas.

Facilitators will find it useful to be very explicit about these guidelines; you might want to post them prominently in the room or generate a handout for participant reference.

Facilitating small- and whole-group discussions

The role of discourse in a *Lenses on Learning* seminar parallels the role of discourse in helping children to make sense of complex ideas in mathematics classes and in helping teachers to make sense of their own teaching practice in professional development programs. Like the students and teachers, administrators may need support and guidance to formulate and express new ideas. They may not realize that in this class, they can expect their ideas to be valued by you and the other participants. Some may need to attend specifically to their own listening skills and learn to actively show an interest in other participants' ways of thinking.

Activities throughout the course involve discussion in either small-group or whole-group configurations. Small-group work allows administrators to explore an issue in depth and uncover its often surprising complexity. For example, in a small-group discussion, people might explore the distinctions between learning "basics" and learning "problem solving." Whole-group discussions often follow the small-group work and can be used to expand on ideas that the small groups have identified as salient. One caution is not to use whole-group discussions simply to report what was discussed in the small groups, which can take up most of the time, but rather to explore the issues further, building on ideas that came up in the small groups.

For flexibility in the use of these course materials, there are often more questions suggested for whole-group discussions than you have time to address. Because of this, one key role of the facilitator is to circulate during small-group discussions, note what issues are "alive" for each group, and then choose from the suggested discussion questions those that will move your group forward on new ideas.

In many cases, administrators will find a discussion especially compelling and may want to talk about the issues for longer than is possible within the constraints of your seminar. Sometimes during a discussion, a particular line of conversation may take over, making it hard for other ideas to be articulated. How do you tell when to stick with a line of conversation and when to move on? More fundamentally, what is the function of these discussions?

All the discussion questions work to bring out the central ideas about mathematics learning and teaching that are embedded in the activity and to connect these ideas to the administrators' professional work. For example, the readings for *Module 1,* Session 1 reveal different perspectives about standards-based mathematics education and give administrators the opportunity to identify and articulate their own current understanding of best practice. Administrators also are asked if they were surprised by anything they read; this question, and others like it that arise throughout the course, are meant to encourage an attitude of curiosity and openness to new ideas.

A good discussion is one in which most of the significant ideas are articulated by someone and extended, reframed, or reacted to by one or more people. The goal is to get the key ideas out on the table and to explore the ramifications of those ideas to some extent. Ideally, you would pursue each idea as far as the group can productively take it at that time. This requires that you make continual judgments about where people are in their thinking and how far they could productively go. When you decide that it is time to move on, you can observe that the subject is very interesting and complex, worthy of more discussion than is possible at this moment. Because the course is designed in such a way that the central ideas come up several times during each module, you can assure the group that they will come back to the same subject at another time.

To effectively facilitate a whole-group discussion, you need to be able to discern the central ideas of the activity and then recognize those ideas when participants raise them. Often, they will express their ideas in halting or undeveloped statements, similar to the ways children discuss mathematical ideas that they are discovering. You will need to listen carefully to what administrators are saying and find ways to extend their points. Some useful strategies are to paraphrase what someone has said, ask the group if they have any reactions to the point, or ask a participant to say more. This will require a bit of improvisation on your part. That is, the discussion questions provided in these materials should elicit the central ideas and concepts, but as the facilitator, you need to listen to the emerging discussion with an ear for "openings"—opportunities to dig deeper—and be ready to go beyond the given discussion questions to elicit meaningful conversation. The goal is not that everyone in the room gets to say everything they think about each reading or activity; rather, the function of discussion is to get administrators thinking deeply about the main issues. When discussions are good, participants will often continue their thinking about a topic later during the session, in their journal writing, after class while they are driving home, or sometime during their workday. The point in class is to help them begin to think about complicated ideas so that they can then carry that thinking into their professional lives.

Sometimes administrators will bring up an issue that is personally quite relevant but is nonetheless tangential to the topic under discussion. For example, while comparing different standards-based curricula in a small group, one participant may comment that none of the curricula being examined is aligned with the standardized tests used in her district. While this is a valid point, to pursue a discussion about the mismatch between standards-based curricula and standardized tests would keep participants from doing what the activity intends: learning to look below the surface features of a curriculum to discover the ideas about mathematics learning and teaching on which the activities are based. As facilitator, you must recognize when the discussions are moving off track. You can acknowledge that the issue being raised is interesting but might be better pursued at another time. Some facilitators keep a running list of such issues on flip-chart paper in a kind of "parking lot," to be discussed when time allows or when they become relevant to the discussion at hand.

Figuring out how much time to spend on a discussion often brings up issues of "coverage" and "closure." Most people carry images of classrooms and teaching in which the teacher's responsibility is to "cover" the text and to come to "closure," so that there are no loose ends at the conclusion of the class. These images are left over from a time in which teaching was fundamentally viewed as transmitting information, and the concern was that students absorb appropriate chunks of information before the class was over. However, today's classrooms are different. They provide structured opportunities for students to think through new concepts, and they assume that for most people, thinking continues long after the class is over. In class sessions, ideas unfold through discourse, and interesting questions may not be completely answered by the end of the class period. In this kind of classroom, it is sufficient to have put important ideas out on the table and to have begun exploring them. The *Lenses on Learning* materials provide a set of activities for each session. However, the goal is not so much to cover all the activities, but rather to have administrators think through a set of ideas. In any given class, you may have to decide whether to let a discussion continue or close it down in order to move on to the next activity. If you choose to move on, the course structure allows you to cycle back to different ideas at later times. To bring closure to a particular activity or discussion, you can offer a few summary remarks that highlight where the discussion has brought the group and what ideas will be revisited at another time.

Many of the ideas that surface in the *Lenses on Learning* course are connected to beliefs that participants may hold very deeply. Some issues will be controversial, and they may be rooted in a range of very different belief systems. Acknowledging these divergent beliefs and talking about why people hold them, rather than trying to reach consensus, will help participants understand both their own and others' ideas.

Someone may feel very strongly, for example, that students should learn basic number facts before engaging in mathematical problem solving, arguing that these facts are tools students need for problem solving. Recent research shows that even without mastery of basic facts, students can solve problems through informal and intuitive ways of working with numbers. Often, they learn the number facts in the process of solving problems and come to understand relationships between numbers as a consequence of solving problems. That is, knowledge of number facts is important, but it does not necessarily have to come first. Nonetheless, the conventional wisdom for many years was that students should learn number facts first, and one or more participants in your class may well hold this belief. If such a participant is to continue to feel like a valued member of the community in your class, you and the other participants need to acknowledge that position as one that many people hold. Indeed, it is valuable for other participants to think through their feelings about this proposition, and how they would talk with someone with this belief, since there are undoubtedly many people who think this way in their communities.

Simply announcing that current research tells us something different will probably not, by itself, change anyone's view. However, over time, as your class does mathematics together and spends time analyzing students' mathematical work, it will be clear that in some cases students know their number facts but do not understand what is going on mathematically in the problem. Through such observations, participants develop a more complex view of the relationship between knowledge of number facts and mathematical understanding.

Setting a climate for viewing videotapes of teachers and students at work

On many occasions in this course, you will be showing a video clip of students describing their mathematical thinking, or of students and teachers working together in a mathematics classroom. The purpose of these videos is to provide concrete images for ideas about mathematics learning and teaching and to offer a starting point for discussions about these ideas. By talking about what they see in the video clips, administrators have yet another context in which they can explore and discuss what mathematical understanding is, how children's mathematical thinking develops, and how instruction can support the development of children's mathematical thinking. The videotapes convey images of elementary mathematics instruction that may not otherwise be readily available to some administrators; from these clips, they become accustomed to what standards-based classrooms look like and begin to recognize meaningful instructional practices.

Although the videotapes were recorded in standards-based classrooms, they are not meant to depict exemplary teaching, but rather, teachers in the process of developing standards-based teaching. Because of this, administrators will see a mixture of very good teaching and teaching that is still under development. This mixture was chosen because, for the foreseeable future, most administrators will see teaching under development when they observe in classrooms.

Facilitators of this course generally find it challenging to elicit intellectually productive discussions of the videotapes. Administrators have been trained to look at classrooms in particular ways, usually in order to assess the adequacy of the teaching and make suggestions for improvement. This viewpoint is understandable, given their responsibilities for classroom observation and teacher supervision. However, in this course, the purpose for viewing and discussing videos is *not* to critique the teaching. Besides, any attempt to do so would not be valid without the benefit of a one-on-one discussion with the teacher. Teaching is very complex; in any situation, there are always other moves that could have been made. Furthermore, the *Lenses on Learning* videotapes contain just short excerpts from a longer class, and participants do not know enough about the background, the context of the lesson, or the teacher's intent to make valid judgments about whether the teacher's decisions were good ones. As one facilitator observed, it is like being given one page from a novel and trying to make a judgment about the novel as a whole.

Everyone should understand that the purpose of the videos in this course is to help administrators develop an eye for today's mathematics classes and to stimulate thoughtful discussion among participants. For example, the video clip *Today's Number* (shown in *Module 2*) depicts a first-grade class talking about different ways to make the number 16. The children have many ways of working with numbers: some count on their fingers; one child uses the number line on the wall; others combine numbers mentally or remember earlier mathematical experiences. This image of a first-grade mathematics class opens up a wide range of ideas: the variety of students' methods for solving the problem (and the fact that in a more traditional form of instruction they might have been taught one "right" way); the easygoing and friendly relationship between the teacher and the children as, together, they explore how the number 16 is composed (and the fact that a more traditional teacher might have seemed more formal and authoritative); and so forth. Participants in your class will likely have different reactions to and interpretations of the teaching and learning depicted in this clip. Such different interpretations are grist for discussion in your class. The issue is not whether a given interpretation is right, but rather, what ideas about mathematics teaching and learning underlie each interpretation.

Someone might say, for example, that the teacher in that video clip should encourage the children who are using their fingers to move on to doing the number work in their heads, instead of validating their method, as the teacher does. Someone else might think it is great that the teacher validates all methods. If two such different viewpoints about the video are voiced in your class, consider yourself lucky! You can use this disagreement to explore the underlying assumptions of each point of view. That is, you could ask the first person, "Why do you think the children should stop using their fingers?" You might hear an answer something like, "Using your fingers is only for beginners, and doing the addition in your head is where first graders should be." This indicates a belief that doing it in your head is a "better" way and that children can learn to do it if the teacher simply teaches the number facts and expects children to use them. Then, you could ask the second person, "Why do you think it's all right that the children are still using their fingers?" Now you might hear, "Moving from using your fingers to counting in your head is a developmental process, and children move through it at different times and at different rates. This classroom is good because it supports the children wherever they are in the developmental process. When children are allowed to develop at their own pace, their understanding will be more solid." Encouraging participants to talk about the reasons they have for their interpretations of the video takes the discussion to a deeper level—it makes the underlying rationale for their thinking clear to others in the class who might see things differently.

When you need to shift a video discussion from an assessment of the adequacy of the teaching to an intellectual discussion about mathematics learning and teaching, ask participants to indicate what specific data in the video support their interpretation. That is, it is more helpful for a participant to say, "The teacher acknowledged that David's use of his fingers was a good way to solve the problem—I felt queasy about that because I think first grade is too late for using fingers," than to say, "This teacher is too touchy-feely and doesn't have high-enough performance standards for the students." The former statement clearly indicates that the participant is making an interpretation of specific data from the videotape. The latter statement is very global, and its grounding in the data is not clear.

As facilitator, you need to help participants make the distinction between what they are learning to see in the videotapes, what they think about what they see, and what they might want to discuss with the teacher in a post-observation conference. This course focuses primarily on the first two of these: helping participants develop an eye for mathematics classrooms and become articulate about the ideas that underlie their interpretations of what they see. (*Module 3* does use one of the video clips as the basis for a supervisor's post-observation conference, but the emphasis here is on recognizing and choosing an appropriate style of communication with the teacher.)

Several specific guidelines will help set a productive climate for eliciting and discussing participants' ideas each time you show a video clip.

- Provide relevant information about the context of the video clip (as provided in these facilitator's materials). Mention the age or grade of students, the teacher's or students' prior experience with the topic, the purpose of the lesson, and other significant background information. This will help set a context and enable viewers to orient themselves to what they will see on the tape.

- Remind viewers that they are seeing brief excerpts from a wider learning context. That is, they are seeing only a small part of a lesson. Thus, they should resist the temptation to make broad judgments about the teacher and the classroom based on this limited view.

- Be clear about the purpose of showing the video: It provides an opportunity to talk about particular ideas. You will generally show each clip twice, once so that participants can take in the events and their sequence, and again in order to focus on particular questions. After the first viewing, ask participants to describe what happened or what they saw; after the second, turn to the questions provided in this book.

- When mathematics thinking is involved in the video clip, give participants the opportunity to work with the problem themselves first, so they are better able to focus on the thinking of teachers and students in the video.

- Remind participants to cite specific evidence from the tape to support interpretations they are making.

- Help participants make the distinction between ideas that are related to developing a discerning eye for mathematics classes and thoughts they have about what would be productive to discuss with a teacher after observing in the classroom.

Examining children's work and teachers' professional portfolios

At several points in this course, administrators will be working with examples of children's mathematical work or excerpts from teachers' portfolios. In all cases, this is the work of real children and teachers. The children's work is in their own handwriting, with an occasional typed transcript included for ease of reading. Here is one example.

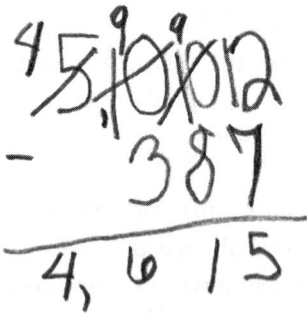

To solve the problem 5,002 – 387, this girl has "borrowed" to make it possible to subtract 7 from 2 in the right-hand column. Her writing makes this clear. In this particular case, the child was being interviewed on videotape and gave an oral explanation; in other cases, participants will see only the written work and will need to infer the child's thought process in order to understand what the child understands mathematically and what he or she is still struggling to understand.

In discussing student work, participants should be encouraged to maintain the same attitude of respect for the child that characterizes the tone of the discussion when watching the videotape. Discussion of student work should take the form of conjectures for which evidence is presented. That is, for the piece of work above, it would not be appropriate to say, "Sasha really is behind for a fifth grader. Most of the kids in my school can do this in their heads and don't have to write it all out." A more useful comment would be something like, "The way she writes out the 4 and the 9s makes me wonder if Sasha understands the meaning of what she is doing. I wonder what question I could ask her that would shed light on that."

Discussing excerpts from teachers' portfolios involves similar considerations. The samples represent the attempt of real teachers to think through new ideas, and the excerpts were chosen specifically to reveal teachers' thinking at moments when they are articulating what is new or hard for them. This helps administrators see what teachers might need to learn afresh, or what they might struggle with, as they move to transform their teaching. For example, one teacher writes:

> *It never occurred to me to ask children how they know that a 10 is ten, or why they carry it over to the next column, in the depth I now know how to do.*

Just as when they look at children's mathematical work, participants looking at teachers' writing should focus their discussion on what the teacher seems to understand and what she is still working to understand, as revealed by the data, rather than making evaluative comments about the teacher.

Discussing administrative practice

As administrators work through the *Lenses on Learning* course, they talk about issues in mathematics education that are being raised in their own schools and districts and consider how some aspects of their own work might shift in order to support the implementation of standards-based mathematics instruction. Throughout, administrators are encouraged to link what they are learning to their own administrative practice. In any one group, circumstances may vary widely. Some administrators may have teachers who have already embarked on the standards-based mathematics agenda, and they may have district policies in place that are moving in this direction. These administrators will have the opportunity to observe new forms of teaching in their own buildings and will likely be seeking information about how to support those teachers and extend the work more broadly in their schools. Other administrators may not have any teachers who are moving toward change and, instead, may be in the position of initiating change themselves. A third group of administrators might be ambivalent about standards-based mathematics education or even disagree with it and value traditional mathematics instruction. Administrators in these different circumstances will likely think very differently about how standards-based mathematics education might affect their own work. The course takes this into account in the design of practice-oriented discussions and homework assignments.

Establishing a community in which administrators can reflect on and discuss ideas related to their work may not be easy. The work of school or district administration, itself, does not include many opportunities for administrators to deliberate together about important educational issues. Administrators' jobs require them to act quickly to resolve complex issues. There is little time for reflection about their work and, often, little appreciation of its benefits. An assistant superintendent who took a *Lenses on Learning* class described it this way:

> *As administrators, it is our nature to want to move into action. [In the course that I took], we focused on the ideas or underlying assumptions in a reading, in a [mathematics] problem, or in a videotape, rather than rushing in with an action plan to solve the problems they presented. . . . The course offered [participants] an unusual and safe haven for grappling with real intellectual and professional change.*

Remember that it may take real work for administrators taking this course to develop a taste for, and skill at, listening for the ideas and issues that are embedded in the discussion, rather than moving immediately into problem-solving mode.

While participants need the chance to step back from the pressure of their everyday responsibilities to participate in and benefit from the reflection on fundamental issues about teaching and learning, it is equally important for them to actively connect the emerging new ideas with their administrative practice. The Bridging to Practice reflective writing at the end of each session asks participants to reflect in a journal entry on the issues that came up for them in the class that day, to consider where they and their schools are regarding these issues, and to think about what they would do to move themselves and their schools along. In this context, "doing" can consist of further personal reflection and learning, as well as more overt action.

In this writing, some administrators may be inclined to give you feedback on the course and how it is working for them. Bridging to Practice is not meant for this purpose. However, recognizing this instinct, you might give people the chance to provide such feedback several times throughout the course. One strategy is to provide "exit cards," index cards on which they can write their reflections just as they leave a session. If you decide to do this, emphasize that they should write substantive, practical comments that you can use. Help them understand that general comments such as "Great course, good discussions!" or "Discussions too long!" are not especially useful to you as facilitator.

Your own journal writing

Keeping a journal about the ideas and issues you encounter while teaching a *Lenses on Learning* course offers opportunity for sustained reflection and growth. In general, your writing should focus on what the administrators in your course are thinking about the ideas that emerged from each session, rather than solely on your own choices as facilitator or on what worked or did not work as a teaching strategy. For example, two facilitators wrote the following in their journals:

> *In viewing the videotape, most people got [what the child was doing]. But the person who really didn't see it (and spent some time trying to explain it quite differently) was the head of a high school math department. She is really grounded in the traditional math . . . and she is really trying to learn. But she might be having a hard time following the "logic" of those video sequences. Perhaps there is a little short circuit that happens to a "math" person who sees things in the traditional way.*

> *One participant brought in an article about the letter from 200 mathematicians, expressing their opposition to the DOE curriculum representations. Because his school is going through a curriculum adoption process, he fears community backlash and is very much concerned about not repeating the mistakes of the whole language/phonics battle or of New Math. He doesn't want to throw the baby out with the bath water, wants to have a balance, etc. I'm appreciating the ways in which he's articulating his concerns. He's clearly interested in standards-based mathematics, aware of his own mathematical limitations, and gradually trying to make sense of it all. I encouraged him to write about his conceptual struggles, but his Bridging to Practice reflective writing is about some of the standards-related ideas that are compelling to him. Interesting fellow to be working with.*

By attending carefully to the thinking of the administrators in your group, focusing sometimes on the group as a whole and sometimes on individuals, you will gradually get a picture of the ideas and strengths that they bring to the class and the ideas that are new to them. With this knowledge, you will be able to build your teaching on participants' thinking, much as teachers in mathematics classrooms are coming to do.

Introduction to Module 2

Teacher Learning for Mathematics Instruction

When the National Council of Teachers of Mathematics published their *Standards* in 1989, there was little information about how teachers might learn to teach in the way envisioned in those *Standards*. Conventional staff development at that time was oriented toward helping teachers assimilate new techniques into an existing system of ideas about pedagogy and subject-matter knowledge. The 1989 NCTM *Standards* and the subsequent *Principles and Standards for School Mathematics* (2000) require not only that teachers master new technical skills, but also, more importantly, that they reconceptualize and reinvent the overall nature of their teaching practice. This requires a fresh approach to professional development for teachers.

What do teachers need to learn in order to reconceptualize their mathematics teaching? Over the past several decades, research has provided new information about how children learn mathematics. This research shows that children actively think about mathematical ideas as part of their ongoing efforts to make sense of the world, even before they enter school. However, classroom instruction has too often paid scant attention to the mathematical understandings that children bring with them and has presented mathematics as a set of facts and procedures, with little examination of the underlying patterns and ideas. In these circumstances, many children lose touch with themselves as mathematical thinkers and think of "school math" as dry and incomprehensible.

Some learners of mathematics are able to retain their sense of themselves as mathematical thinkers and make sense of the ideas embedded in the formal symbols and rules on their own. Many others need a teacher's help to make the link between their own mathematical ideas and those underlying the formal school curriculum. Otherwise, they may experience mathematics as the memorization of facts and formulas, without meaning.

Fundamentally, standards-based mathematics aims to harness the sense-making instinct in children. To be sure, it is important for students to learn the facts of the number system and how to do basic calculations; these skills contribute to mathematical fluency and efficiency. Standards-based mathematical instruction, however, opens the door to the learning of powerful ideas as well. The teacher's task in this process is to discern the sense that children are making of mathematical ideas and to connect their sense-making with the school curriculum. This idea is new to many teachers.

Most teachers were trained to "stand and deliver" the facts and skills of their subject. When they were encouraged to listen to students, the purpose was primarily to affirm children's right answers and to correct their errors. In order to exploit children's sense-making energies and turn students into powerful mathematical thinkers, teachers need to develop very different ideas about the process of learning, or sense-making, and about the process of teaching. They need to examine how children come to understand the major topics in K–12 mathematics. At the same time, they need to increase their own mathematical knowledge so that they can interpret children's ideas and make productive judgments about where to guide the children next.

Essentially, then, teacher learning through professional development needs to encompass three things:

- **The opportunity to deepen their own mathematical knowledge** For teachers, this is not simply a matter of taking more mathematics courses. Rather, they need to work through fundamental mathematical ideas at both concrete and abstract levels, exploring ambiguities, and persisting with their puzzlement until the mathematical ideas really make sense to them.

- **The opportunity to develop a rich understanding of how children come to understand the major topics in K–12 mathematics** This understanding comes to teachers as they read research-based articles on children's mathematics learning, closely examine student work, learn to attend to children's mathematical reasoning, and consider where and how children's ideas connect with the mathematical concepts in the curriculum.

- **The opportunity to think deeply about the nature of learning** Once again, this is not simply a matter of learning new facts from educational research. Teachers must reconsider long-standing beliefs about learning. They must rethink the old notions that learners are empty vessels waiting to be filled; that students learn by being told what to do and how to do it; that mathematics consists of a series of isolated facts and topics that should be taught in a certain order; and that students' confusion should be relieved by the teacher, rather than interpreted as a sign of important thinking.

Lenses on Learning, Module 2 gives administrators the chance to explore how they, as instructional leaders, can support teacher learning in these areas.

At the same time, administrators must take into account how the profession of teaching is changing: Teachers who once worked as isolated experts in their classrooms are now working as colleagues who inquire together about the nature of children's learning and good teaching. Being part of a community of professional adults who are involved in this collaborative inquiry is now seen as central to teachers' growth. Administrators will need to consider the implications of this for teachers' professional development.

The new ideas about learning, about teaching, and about professional development that are addressed in the *Lenses on Learning* course have been emerging into the mainstream of educational thought in the years since most administrators were trained. In fact, much of the research on teaching that took place in the 1970s (termed *process-product* research), which has informed much of the thought about teaching and learning since then, focused not on teachers' and students' thinking about mathematical ideas but, rather, on the kind of teacher behaviors and teaching techniques that were statistically associated with high performance on standardized tests—such things as employing "wait time" and "higher-order questioning," having a clear structure to the lesson, and making clear transitions from one activity to another. Many professional development programs on effective teaching and on teacher supervision have focused on such generic features of teaching—specific behaviors and techniques that can be easily identified and learned. While these techniques are useful tools for teachers as they work across disciplines, still more is required of teachers when the goal is to deepen and expand children's mathematical knowledge.

Some of the administrators in your group may not yet recognize the shift that has occurred in the nature of professional development for teachers, especially with respect to mathematics. Other aspects of teacher learning may be new to administrators as well: the amount of time it will take for teachers to make the changes called for in their teaching practice; the kinds of professional development programs that are likely to be of help; the type of support that teachers will need; and the nature of administrators' own roles as instructional leaders. This last aspect may be particularly challenging to administrators. Many may be unprepared for the degree of collaboration required between them and teachers as they plan and provide a range of professional development opportunities in their schools and district.

During their work in this module, administrators think about all these issues. The module unfolds in five class sessions.

- ◆ **SESSION 1 Changing More Than One's Socks** Administrators explore the range of ideas that teachers need to consider as they reconstruct their teaching to align with NCTM's *Principles and Standards for School Mathematics,* with current frameworks, and with new mathematics curricula. The big ideas explored in this session include

  - the beliefs, knowledge, and pedagogy that teachers need to develop in order to fundamentally change their mathematics teaching practice
  - the kind of support that teachers need in order to make such changes
  - the time frame required for making such changes

- ◆ **SESSION 2 What Do Teachers Need to Learn?** This session focuses on two important areas of teacher learning: deepening their understanding of mathematics itself, and learning to listen to children's mathematical thinking in order to understand how their mathematical ideas are developing. Administrators specifically consider the strengths and needs of teachers in their own schools in each of these areas. The big ideas explored in this session include

  - the need for teachers to understand the mathematics they teach
  - the need for teachers to cultivate their ability to listen to students' mathematical thinking and reasoning

- the characteristics of professional development experiences that could help facilitate teacher learning

♦ **SESSION 3 What Makes for Meaningful Professional Development?** This session begins with a look at another central aspect of teacher learning: how to facilitate rich mathematical discourse in the classroom. Administrators make visual representations that profile the teachers in their schools to better assess the specific professional development needs they face. While viewing a videotape of teachers who are working together to learn mathematics, participants begin to identify characteristics of meaningful professional development. The big ideas explored in this session include

- the role of classroom discourse in mathematical learning
- the varying professional development needs of teachers within any school or district
- characteristics of professional development that fosters learning to facilitate discourse

♦ **SESSION 4 Critical Colleagueship** This session introduces several alternative images of professional development. Administrators consider both the nature of these learning opportunities and the school culture that is created. They explore critical colleagueship—the idea that teachers can (and should) work together in substantive and intellectually critical examinations of learning and teaching. To further develop their perspective on characteristics of meaningful professional development, participants examine several models and approaches. The big ideas explored in this session include

- the differences between the training paradigm and critical colleagueship as alternative approaches to professional development
- different kinds of professional development that build on critical colleagueship
- the promises and tensions inherent in professional development that embodies critical colleagueship

♦ **SESSION 5 Providing Professional Development** Participants now shift their thinking from the ways professional development can meet *individual* needs to the ways in which these experiences can build on each other within a larger school or district community. They reflect on the tensions inherent in their twofold task: responding to individual needs and managing systemic work. The big ideas explored in this session include

- the relationship between the learning a teacher needs or wants to do and the range of professional experiences that might provide that learning
- the promises and tensions that emerge when trying to meet the needs of individual growth and, at the same time, move entire systems forward

Preparation for the seminar

Well in advance of the first class session, facilitators should send each participant the readings for *Module 2* along with Handout 1, the preparatory homework assignment. Before the first class, participants are expected to read and think about the first reading.

As facilitator, you will need to study Reading 1 yourself and be prepared to draw out participants as they discuss the sections of the article that were of greatest interest to them.

Reading 1 is fairly dense, more so than others that are assigned in this module. It was chosen because it brings together several important aspects of change that standards-based mathematics asks of teachers. A further intent was to include a reading that reflects a research orientation to professional development. Despite the article's density, the discussions it has sparked with administrators have been fruitful ones.

Homework in Preparation for *Lenses on Learning, Module 2*

You have enrolled in the course *Lenses on Learning, Module 2: Teacher Learning for Mathematics Instruction*. In preparation for our first session, please do the following homework.

With this homework assignment, you should have received a book with all the readings for *Lenses on Learning, Module 2*. First, read the book's brief introduction to get a sense of what to expect from the course. Then read the first article.

READING 1 "Characteristics of a Model for the Development of Mathematics Teaching" by Lynn T. Goldsmith and Deborah Schifter

We will discuss the article in our first class meeting. As you read, please mark those paragraphs or short sections that seem most salient to you. You will be asked to share one of your selections with other course participants. Be prepared to draw out the central points of that selection and to explain why it was of particular interest to you.

SESSION 1

Changing More Than One's Socks

Teachers are being called upon to reexamine their assumptions and to change their views in order to align their practice with new beliefs about mathematics teaching, learning, and assessment. The breadth and depth of these changes require us to consider new ideas about professional development for teachers. That is, professional development needs to move beyond the acquisition of new teaching techniques, to encompass a deepening of teachers' understanding of mathematics as well as the nature of children's mathematical thinking. Such understanding is essential if teachers are to listen productively to children's mathematical thinking and make good decisions about how best to support the development of their students' ideas.

Professional development for teachers has traditionally been seen as "one size fits all," but in fact there is wide variation in teachers' mathematical knowledge and their understanding of how children's mathematical knowledge develops. Many administrators do not realize that teachers in their schools may well need different kinds of learning experiences. Furthermore, many administrators are accustomed to providing professional development experiences of short-term duration and are not attuned to the time needed to make these new changes. There are no quick fixes; the comprehensive nature of the professional development experiences considered here demand a long timeline and ongoing support.

Administrators, then, need to explore several questions as they consider new ways of supporting teacher learning.

- Just what are these new beliefs about learning and teaching that teachers need to consider?

- What is the nature of the mathematics knowledge that teachers need if they are to listen knowledgeably to children's mathematical thinking?

- How can administrators tell what issues a teacher is grappling with in his or her own learning?

- How long will it take for substantive change to become evident in teachers' practice?

OVERVIEW FOR SESSION 1

OPENING page 32 30 minutes	**WELCOME** As the group assembles for *Lenses on Learning, Module 2*, introductions are made and administrators learn what they can expect from this session and from the entire module.
ACTIVITY 1 Readings page 34 65 minutes	**DIMENSIONS OF CHANGE** In a whole-group discussion, administrators consider Reading 1, "Characteristics of a Model for the Development of Mathematics Teaching," which describes the process of rebuilding one's teaching practice around new ideas about mathematics teaching and learning. This discussion may reveal a range of viewpoints as participants share their beliefs about the changes in mathematics education and the support teachers need as they make changes in their practice.
ACTIVITY 2 Discussion page 37 70 minutes	**WHAT ARE TEACHERS GRAPPLING WITH?** By reading and discussing a set of statements made by teachers who are exploring new pedagogical approaches, administrators have the chance to hear some of the ideas in Reading 1 expressed in teachers' own voices. As they discuss possible support for these teachers, they begin to think about the challenges they face in their own schools or districts.
CLOSING page 41 15 minutes	**BRIDGING TO PRACTICE** Participants finish the session with guided reflective writing that helps them link the ideas in Session 1 to their own work as administrators. **HOMEWORK** The assignment is a reading that examines the collegial relationship between a professor of teacher education and an elementary school teacher who is working to change her approach to teaching mathematics.

BIG IDEAS	This session explores
• the beliefs, knowledge, and pedagogy that teachers need to develop in order to fundamentally change their mathematics teaching practice
• the kind of support that teachers need in order to make such changes
• the time frame required for making such changes
MATERIALS	☐ Flip chart and markers
☐ Reading 1 for *Lenses on Learning, Module 2*
☐ Quotation strips, pp. 43–44
☐ Handouts 2–3, pp. 45–46
☐ Bridging to Practice display chart or overhead transparency (see below)
PREPARATION	• Prepare and post an agenda for the session that lists the activities and approximate time allotted for each.
• Read through all the materials for Session 1, both the activities and facilitator's notes.
• Review Reading 1, "Characteristics of a Model for the Development of Mathematics Teaching." Make note of the paragraphs or short sections that especially strike you; this will put you in touch with your own thinking about professional development for teachers. If you have worked with the same group of administrators in other *Lenses on Learning* modules, take stock of the diversity of values and understanding that they have expressed so far. What impact have these values and understandings had on the group's progress? You may want to anticipate likely reactions to particular passages in Reading 1.
• Duplicate and cut apart the quotation strips (pp. 43–44) so that you can give a single teacher quotation to each small group. Read and consider these quotations for yourself, as if you were a participant. That is, (1) identify the issues that each teacher is struggling with, and (2) think about the questions you would have in figuring out how to support each teacher.
• Prepare a display chart or overhead transparency of the three writing prompts for the Bridging to Practice reflective writing that participants do at the end of each session (page 41). If you have presented other modules of this course, you may already have a display prepared. Save this display for use in each session.

OPENING
30 minutes

> **Materials**
> Posted agenda

Welcome

20 minutes | **Introductions** Since this is not the first *Lenses on Learning* module, many of the administrators in your group may have worked together before. However, there may also be some newcomers. Spend time on introductions, as appropriate.

Once everyone has assembled, invite participants to take turns mentioning one aspect of the professional development of teachers in mathematics that is especially relevant to them and their district right now. This will give you a sense of where people in the group are starting from. It will also help participants learn a bit about the thinking and the practical circumstances of their colleagues.

Use a flip chart to keep track of ideas they raise. This can be kept as a record of participants' initial ideas.

10 minutes | **Establishing group norms** Let participants know what will be expected of them for this module. Discuss meeting schedules, the nature of the homework assignments, and anything else that seems important to the circumstances of your group. If any participants are new to the *Lenses on Learning* course, explain the idea of establishing a reflective community of inquiry within the group.

Review the posted agenda as a preview of this first session.

FACILITATOR'S NOTES
OPENING

Introductions

Participants bring to the course particular personal perspectives and sets of interests drawn from their own prior experiences, cultures, and current work in schools or districts. Encouraging group members to talk about themselves and why they are taking the course helps everyone to learn a little about each other and glimpse the range of professional responsibilities, ideas, resources, and concerns represented in the group.

Giving participants a chance to talk about their own priorities and struggles will communicate the fact that you, the facilitator, understand and acknowledge the complexity of their professional realities. By the same token, it can be valuable to communicate that you, too, are a learner in this endeavor. By sharing a recent understanding that you have reached, you can convey your stance as a peer explorer, though perhaps one who has started on the journey earlier than many of them. In addition to releasing you from the role of "expert," this tells participants that their own journey could be an extended one, too.

Establishing group norms

Let participants know that articulating their own thinking and reflecting on the thinking of others are essential to the success of this course. Everyone must be attuned from the start to building supportive relationships with their colleagues in the seminar. This involves not only listening attentively to others, but also being willing to take the personal risk of sharing thoughts that are not yet fully developed and expressing their own questions and confusion.

Emphasize that everything said in the room will be kept confidential. This frees participants to explore their thinking openly, without concern about their vulnerable spots being exposed in wider professional circles.

Reinforce the importance of regular attendance, prompt starting times, and completion of homework between sessions. If everyone is committed to the course, administrators can expect their discussions to grow deeper and richer over time.

ACTIVITY 1
Readings
65 minutes

Materials
Reading 1
Flip chart and markers

Dimensions of Change

65 minutes

Whole-group discussion Before this session, administrators will have considered Reading 1, "Characteristics of a Model for the Development of Mathematics Teaching," and marked those sections that interested them. Structure the discussion so that each section a participant brings up is explored at some depth before the discussion moves to the next one. You might start this way:

Who would like to read to the group a section that you found especially interesting? Tell us the page number so that we can read along. After reading, please draw out the central points and tell us why that section was interesting to you.

Invite comments from the group by asking, for example,

What do others think of this section? Did anyone else highlight this one? Why? Do people agree with this interpretation?

When discussion of the first passage is finished, ask if anyone else picked out a similar passage related to the same topic. Ask each subsequent volunteer to read, summarize, and comment on the section he or she chose, then invite further discussion from the group. When there are no more passages related to the initial topic, ask if anyone was intrigued by a different set of ideas in the article. Continue as before. As the discussion moves along, keep track of the main ideas on the flip chart.

THE BIG PICTURE
The goal here is to make clear the difference between professional development that invites teachers to consider a range of ideas and an approach that instructs teachers to adopt a new set of strategies. Standards-based mathematics education requires that teachers change their beliefs about the nature of mathematics, children's development of mathematical ideas, and how classroom learning of mathematics can take place. From this reading, administrators should become aware of the complexity of this kind of teacher learning. Such learning necessarily takes place over a period of years, and thus should be considered a long-term investment.

TRANSITION

Explain that in the next activity, participants will read actual quotations from teachers grappling with the process of change. These quotes will help to ground the ideas from the reading they just discussed.

FACILITATOR'S NOTES
ACTIVITY 1

Whole-group discussion

Reading 1 brings together several distinct and interrelated aspects of change that standards-based mathematics education asks of teachers: the need to deepen their mathematics knowledge, the need to change their beliefs about the nature of learning, and the need to change their classroom instructional practices.

In this discussion, participants express their *current* perspectives. Note the emphasis on *current,* which suggests that beliefs can evolve with experience and reflection. In a heterogeneous group, you are likely to hear a wide range of views. Some people will think of professional development mostly as training in new techniques; others will recognize that professional development can support changes in underlying beliefs. Letting this rich discussion play out without trying to resolve who is "right" gives participants the chance to hear ideas that are new to them from peers they respect.

Do not be surprised by administrators who come to this course with the idea that professional development primarily entails learning new teaching strategies and techniques. One goal of this discussion is to convey that professional development for teachers entails much more than the acquisition of new instructional practices. Participants may highlight paragraphs that point to the foundational nature of this change: "We're talking about big shifts, and not just the new look." They may emphasize the longer time involved in teachers making such changes. Some may point to the idea that teachers who are beginning to change their practice may demonstrate behaviors that reflect a mixture of older and newer sets of beliefs.

♦ **The facilitator's role** As the facilitator, you are responsible for ensuring that the central points in the reading are raised and identified. Use your own highlighted paragraphs as guidelines. If, for example, no one has mentioned the importance of teachers understanding the mathematics for themselves, then you might explicitly ask if anyone has marked this idea.

To encourage dialogue among participants, point out connections among the paragraphs they select. For example, as you list the points on a flip chart, try organizing them into themes. You might ask the group what broader themes they see emerging from their discussion. This can help participants see how their individual ideas contribute to building a deeper understanding of the complexity of mathematics education.

♦ **Dealing with differing opinions** The discussion may raise some differences of opinion about best practice in mathematics education. Not all participants will agree on the relative merits of giving students opportunities to construct their own understanding of mathematical concepts versus giving them procedures for solving problems.

You may also hear differing beliefs about the role of confusion in a mathematics class. Some participants may feel that confusion is to be avoided and should be cleared up right away by the teacher so that it does not interfere with understanding. Participants who take this stance may express concern that confusion is frustrating to students and could turn them off to mathematics. On the other side, participants may see confusion as a productive phase in the learning process. They may argue that teachers need to encourage confused students to further explain their ideas; only then can the teacher work to move these students beyond their confusion. Rather than prolonging any argument, explain that confusion is a complex issue that cannot be reduced to simple statements such as "confusion is an important learning phase" or "confusion only inhibits learning."

Highlight intellectual tensions as they arise and note that these issues will be revisited throughout the module. This is especially important when the differences stem from deeply held beliefs. When such differences arise, look for elements they have in common—such as a genuine concern for children's motivation to learn. Ask participants to look for the assumptions about learning and teaching that underlie their beliefs, and to consider how those beliefs have led to such differences of opinion in spite of a shared common concern for children's learning.

Do not push the group to come to a consensus on a difference; rather, aim for mutual understanding of the different positions. In order to build a trusting community, participants need to learn how to express differences without feeling at risk.

◆ Connecting ideas to the workplace

Administrators have likely begun thinking about the implications of this reading for their own work. Some may simply recognize that teachers need support and help in their efforts to develop their own practice. Others may feel overwhelmed by the amount of work to be done. As one participant wrote,

> *Sometimes I feel that I am a little ant—I can do so little, and the mountain seems so high! I wish I had a way to measure—keep track of progress, movement—though sometimes I'm not sure whether that would make me feel encouraged or discouraged!*

Some participants may wonder about the extent to which they should help teachers who are struggling with new ideas and practices. As one asked himself,

> *In working with teachers when they are struggling with something new, how much help do you give? To what degree do you let them struggle?*

In districts where teachers come from different cultural backgrounds or have different frameworks for mathematical education, administrators may be concerned about how to work with such a diverse group. A challenge these administrators face is how to respect teachers' beliefs and frameworks while still holding out an expectation for change. For instance, one teacher may believe strongly that students should not question teachers. An administrator would have to figure out if this view has a cultural basis, emanates from a pedagogical orientation, or is rooted in a teacher's mathematical insecurity.

When participants find the practical implications of Reading 1 overwhelming, you will want to be especially supportive and understanding. Reassure them that while much of this module focuses on *what is needed,* a substantial portion of it will shed light on *how to help* teachers.

ACTIVITY 2
Discussion
70 minutes

Materials
Quotation Strips (cut from pp. 43–44)
Handout 2
Flip chart and markers

What Are Teachers Grappling With?

20 minutes | **Small-group discussion of teacher quotations** Break the class into several small groups of three or four people each. Give each group a quotation strip.

Explain that these statements are adapted from comments made by teachers in a professional development program who were in the process of rethinking their own mathematics teaching.

Give small groups about 20 minutes to discuss their quotation, answering the two questions on each strip:

What do you hear this teacher thinking about or grappling with?

What kind of support might best help this individual to grow as a teacher of mathematics?

Ask each group to choose a reporter who will share the group's thinking with everyone in a whole-group discussion.

To help participants develop an appropriate orientation to the small-group work, you might say the following:

Think about these statements not from the perspective of traditional "supervisors," but rather as administrators who have responsibilty for providing instructional leadership within the school.

Rather than forming opinions about what each teacher said, set your sights on what these teachers need to learn about mathematics, about learning, about teaching, and about assessment of student learning.

Expect that people in your group may have different ideas about some of these issues. Rather than being problematic, these differences can lead to a richer discussion.

As you listen to the small-group discussions of the quotes, take note of the different interpretations and recommendations you hear. You can bring up these differences in the whole-group discussion that follows.

25 minutes	**Sharing interpretations** Call participants back into the large group configuration. Distribute Handout 2, *Collected Teacher Quotations,* so that everyone can read them all.

Working with one quotation at a time, ask group reporters to say what their group thought the teacher was struggling with and to give their ideas for supporting the teacher's professional growth.

In this discussion, listen for the ways participants talk about working with teachers. To what extent do they appear eager to help teachers construct their own understanding? Are some individuals assuming a "delivery" approach to professional development? You might point out and seek reactions to what you observe. |
| 25 minutes | **Supporting teachers** Open the discussion to include related issues administrators in the group face in their own work, asking,

What issues do you face in your school or district as you seek to support teachers?

List these issues on flip-chart paper that you can keep for reference in Session 5 of this module.

THE BIG PICTURE
This open discussion allows administrators in the group to hear what aspects of professional development for teachers are most on the minds of their colleagues; to connect the ideas of the seminar to their own practice; to see that there are many hard issues to deal with; to remember that no one has all the answers; and to discover that thinking about problems together can be helpful. It can provide you, as facilitator, insights into participants' understanding of and concerns about the topic of professional development for teachers.

If participants begin moving the discussion to ways to "fix" these problems, mention that this is not the time to try solving the problems they have identified. When building a community of people who can inquire together about important issues, discussions like this must not become predominantly discussions of "things that work," but rather inquiries into what makes these issues important and hard. |
| TRANSITION | Tell participants that in the next session, they will explore two important areas of teacher learning: deepening their understanding of mathematics and learning to listen to children's mathematical thinking. The homework will provide grounding for this exploration. |

Facilitator's Notes
Activity 2

Small-group discussion of teacher quotations

The quotations are taken from the portfolios of teachers who were participants in a multiyear professional development program. The quotes lend themselves to different interpretations and a variety of practical responses. (Refer to About the Quotations, below, for further information.) However, each illustrates general issues that typically arise for teachers who are exploring new pedagogical approaches. Discussing these quotes and listening for the different ways in which people interpret them is good practice for participants' work with teachers and other administrators in their own schools and districts.

About the Quotations

Teacher A

This teacher is coming to recognize that knowing what to do in class at any given time comes from paying careful attention in order to gauge the level of students' understanding. As Goldsmith and Schifter write, "student understanding can be both the guide and the goal of [teachers'] practice." Teacher A is also expressing the new realization that the answer to the question of what to do is not found in a book. Rather, by listening to the mathematical thinking of students, this teacher can find the knowledge sought.

To support such a teacher, an administrator might encourage him or her to get together with other teachers so they can pool their growing understanding of their students' mathematical thinking.

Teacher B

This teacher has come to recognize the critical importance of a teacher's understanding of the mathematical ideas. He observes that creating one's own understanding of a concept is much more meaningful than following someone else's explanation. In light of the success this teacher experienced as a learner in constructing his own understanding, he now realizes that simply providing explanations to his students is no longer meaningful.

To support the growth of this teacher, an administrator might provide more contexts in which the teacher has the opportunity to construct his understanding of important mathematical concepts.

Teacher C

This teacher is reflecting on the limitations of learning mathematics procedurally. She admits that students' ability to repeat what she said in their own words was something she mistook for true understanding. Real understanding is lasting, she realizes—confirming that her students' apparent understanding, which was so short-lived, was not the real thing.

To support this teacher's professional growth, an administrator might provide her with opportunities to construct her own understanding of mathematical concepts, or offer a curriculum that structures experiences so that students can construct their own understandings.

Teacher D

This excerpt from a teacher's journal highlights the emotional difficulties teachers may face as they undertake major changes in their teaching practice. Even with the help of a mathematics coach, the whole enterprise can seem overwhelming.

For support, an administrator might put such a teacher in touch with other teachers who are embarking on the same journey. Additional course work might also help this teacher. Another good support might be a curriculum that specifically helps teachers understand the mathematics underlying each activity or problem, explains how students might be thinking about the mathematics, and suggests ways to work with the range of students' mathematical ideas.

Sharing interpretations

While thinking about an individual teacher and the kind of support that might help that person grow as a teacher of mathematics, administrators practice generating specific administrative actions or responses to teachers who are at different stages in their professional development.

As you facilitate the whole-group discussion, respond to administrators' suggestions with questions that probe their reasons or rationale for particular actions. For example, you might ask them to think about the *consequences* of taking certain steps: What do they hope the teacher would do? What do they think might really happen? In this way, you are helping the administrators to envision, in a very concrete and practical sense, potential consequences of their actions.

As each group shares its interpretation of one specific quote, you might actively solicit differing perspectives. If you overheard small-group discussions in which participants had different views, encourage them to air these. For example, you might say,

> *When I was listening to the group discussing the quotation from Teacher D, about the panic attack, I heard that some people were excited and encouraged while others were worried and concerned. Can you say a little about why you had these different reactions?*

Remember, there is no need to resolve such differences and, in fact, it would probably be counterproductive to do so. Rather, underscore that both sets of ideas are reasonable interpretations. Making them explicit can help bring to light the different assumptions that administrators may have about learning, teaching, and professional development.

Keep in mind that your own actions or responses to what participants say can provide a model, or an example, of what can happen between administrators and teachers as well as between teachers and students.

Supporting teachers

This discussion turns administrators' attention to the professional development of teachers they work with in their own settings. You may hear a range of issues. Some administrators may be puzzling about how to work with "reluctant" teachers—those who think their current teaching styles are just fine, or who have reservations about starting something new. Others may wonder how they can help teachers feel safe as they push the boundaries of their knowledge and practice. Some may be wondering how to make available to teachers the kind of time they need to undertake significant change; they may also wonder about the financial implications of such support. Still others may be thinking about how they can provide instructional leadership that respects and encourages teachers' professional autonomy.

Again, discourage any discussion of ways to "fix" these problems, but reassure administrators that in subsequent sessions they will be exploring these topics further and considering professional development that will address diverse needs.

The list you make of the specific challenges that participants identify in their own situations will be used again in Session 5.

CLOSING
15 minutes

Bridging to Practice

Participants who have experience in another module will be familiar with the Bridging to Practice exercise. For those who are new to the *Lenses on Learning* course, explain that the purpose of this reflective writing is to enable them to focus on the ideas in the course that seem most relevant to them and to the school community in which they work. Bridging to Practice provides a structure for planning practical actions they could take in their work situations.

Display the three standard Bridging to Practice questions:

1. Pick an idea that came up today that you found particularly interesting. What is your current thinking about this idea?

2. Where is your school now with regard to this idea?

3. What are one or two things that you will go back and pursue, to move yourself and/or your school along with this idea?

Homework

For homework before Session 2, administrators need to consider Reading 2, "Learning to Hear Voices: Inventing a New Pedagogy of Teacher Education," by Ruth Heaton and Magdalene Lampert. This reading is excerpted from an article that describes Heaton's experience (as an elementary teacher) as she worked with Lampert (a teacher educator) to transform her teaching. Heaton's description of her response to the challenges puts into a more personal perspective many of the ideas described in Reading 1.

Reading 2 prompts administrators to think further about the ideas that were discussed in the first session. At the same time, it orients them toward the second session, in which they will think further about what teachers need to learn as they seek to align their teaching with the ideas of mathematics reform.

Facilitator's Notes
Closing

Bridging to Practice

At the end of each session, participants spend 15 minutes on reflective writing about ideas from the session that were particularly meaningful to them, explaining how these ideas relate to their current work. The purpose of the Bridging to Practice exercise is to give participants time to reflect on the ideas discussed in the class and to articulate, for themselves, what is important to them about these ideas. It also helps them make the bridge between these ideas and their own work as administrators.

On the job, school administrators have constant demands on their time and energy. For this course, they must momentarily forgo their daily commitments in order to dig deeply into the new ideas. However, at some point they need to think through how they might carry these ideas into their own practical work as administrators.

Not only will this reflection help administrators connect new ideas to their practice, but thinking about new ideas in the context of practice may shed new light on the ideas themselves.

If this is the first time some participants have engaged in reflective writing, it will be helpful to take a few minutes to talk about the process. Emphasize that the writing is exclusively for their benefit. This is not the place to tell you what they liked about the class.

Whether or not you ask participants to turn in this writing is up to you; you may want to collect copies of their reflections for your own use in understanding the thinking of the administrators in your class. In this case, they may submit their writing anonymously if they prefer.

Your own journal writing

Within one or two days of teaching this class session, you might set aside some time to write your own journal entry. Try to write about the ideas held by the administrators in your group rather than about your own actions and how things worked out. Using the time to jot down what the participants in your class are thinking about will be much more useful to you as you begin to prepare for the next class session. Some facilitators make preliminary notes for this journal writing during the time when administrators are doing their own reflective writing.

Lenses on Learning, Module 2

QUOTATION STRIP
FOR SMALL-GROUP WORK

TEACHER A

Suppose a teacher said this to you in a conversation:

A year ago, I was in search of the perfect resource book to tell me what to do. I looked for answers someplace else. Now, instead, I think, "What are my students thinking about? What interests the range of second graders that I have? What concepts are different students ready to explore?"

DISCUSSION QUESTIONS

- What do you hear this teacher thinking about or grappling with?
- What kind of support might best help this individual to grow as a teacher of mathematics?

Lenses on Learning, Module 2

QUOTATION STRIP
FOR SMALL-GROUP WORK

TEACHER B

Suppose a teacher said this to you in a conversation:

I discovered the meaning behind the operations I had memorized and forgotten many years ago. I experienced what it feels like when math makes sense. I had created my own understanding. If understanding meant more than being able to follow someone else's explanation, it naturally followed that my success as a teacher could not be measured by my ability to explain what I understood to each and every one of my students.

DISCUSSION QUESTIONS

- What do you hear this teacher thinking about or grappling with?
- What kind of support might best help this individual to grow as a teacher of mathematics?

Session 1, *Changing More Than One's Socks* ♦ 43

Lenses on Learning, MODULE 2

QUOTATION STRIP
FOR SMALL-GROUP WORK

TEACHER C

Suppose a teacher said this to you in a conversation:

There were too many times that children seemed to follow the motions but did not retain anything. I think they really didn't know and understand the things I assumed they did. There were usually several students who could repeat what I said in their own words. I thought this meant understanding. I could never figure out why their understanding didn't seem to last very long.

DISCUSSION QUESTIONS

- What do you hear this teacher thinking about or grappling with?
- What kind of support might best help this individual to grow as a teacher of mathematics?

Lenses on Learning, MODULE 2

QUOTATION STRIP
FOR SMALL-GROUP WORK

TEACHER D

Suppose a teacher said this to you in a conversation:

It is now 5 a.m., and I woke up with a panic attack. How am I ever going to be able to do this in my classroom? Day in and day out for 180 days? I guess with the help of the new math coach, "I shall overcome," and after bumbling my way through the first year of it, maybe it will become more of a natural thing. But I am still dreading the weekly "observations" (in spite of the math coach's assurance that it is a "consultation" and not an "observation").

DISCUSSION QUESTIONS

- What do you hear this teacher thinking about or grappling with?
- What kind of support might best help this individual to grow as a teacher of mathematics?

Collected Teacher Quotations

TEACHER A

A year ago, I was in search of the perfect resource book to tell me what to do. I looked for answers someplace else. Now, instead, I think, "What are my students thinking about? What interests the range of second graders that I have? What concepts are different students ready to explore?"

TEACHER B

I discovered the meaning behind the operations I had memorized and forgotten many years ago. I experienced what it feels like when math makes sense. I had created my own understanding. If understanding meant more than being able to follow someone else's explanation, it naturally followed that my success as a teacher could not be measured by my ability to explain what I understood to each and every one of my students.

TEACHER C

There were too many times that children seemed to follow the motions but did not retain anything. I think they really didn't know and understand the things I assumed they did. There were usually several students who could repeat what I said in their own words. I thought this meant understanding. I could never figure out why their understanding didn't seem to last very long.

TEACHER D

It is now 5 a.m., and I woke up with a panic attack. How am I ever going to be able to do this in my classroom? Day in and day out for 180 days? I guess with the help of the new math coach, "I shall overcome," and after bumbling my way through the first year of it, maybe it will become more of a natural thing. But I am still dreading the weekly "observations" (in spite of the math coach's assurance that it is a "consultation" and not an "observation").

DISCUSSION QUESTIONS

- What do you hear each teacher thinking about or grappling with?
- What kind of support might best help each individual to grow as a teacher of mathematics?

Homework for Session 1

Complete the following reading before the next session.

READING 2 "Learning to Hear Voices: Inventing a New Pedagogy of Teacher Education" by Ruth Heaton and Magdalene Lampert

Think about and be prepared to discuss the following questions. Consider taking notes on the reading to use as reference in that discussion.

- How did Heaton's pedagogical beliefs shift?
- What emotional reactions did Heaton have to the things she was learning?
- What is it about Lampert and Heaton as individuals that allowed them to establish such a productive relationship?

SESSION 2

What Do Teachers Need to Learn?

Standards-based mathematics demands much more of teachers than merely incorporating a new set of strategies into their teaching. The goals of mathematics instruction now include helping children to build on and strengthen their *intuitive* mathematical thinking so that they will have a strong *conceptual* understanding of mathematical ideas. As a result, teachers need to be focused more than ever before on both the mathematical ideas themselves and the ways children think about them. Among other things, teachers need

- to understand for themselves, in a deep and flexible way, the mathematics that they will be teaching
- to know how to "hear" and make sense of the mathematical ideas that are embedded in children's talk and written work and to decide what instructional moves to make to extend children's thinking
- to be able to facilitate classroom discussions that further the development of children's mathematical thinking

This session helps administrators develop a framework for understanding the first two of these key areas of teaching practice: understanding the mathematics and interpreting children's thinking about mathematical ideas. The third key area, facilitating worthwhile discussions, is discussed in the next session. Administrators begin to explore why these things are important, how teachers can learn to do them, and why they are hard to learn. This session is not intended to present models of classroom practice or professional development. Rather, the focus is on the kind of learning required of teachers. The intent is to help administrators gain greater insight into the issues teachers are grappling with as they struggle to change their practice.

Overview for Session 2

OPENING page 50 5 minutes	**INTRODUCING THE SESSION** Begin with announcements and a brief preview of the session agenda.
ACTIVITY 1 Readings page 51 30 minutes	**ONE PROFESSIONAL DEVELOPMENT EXPERIENCE** Administrators discuss Reading 2, "Learning to Hear Voices: Inventing a New Pedagogy of Teacher Education." This reading presents the experience of teacher Ruth Heaton, who is working to change her mathematics teaching with the help of a professor of education. Discussion of the reading helps administrators reconnect to the ideas introduced in the previous session and to see them through the eyes of a teacher–learner.
ACTIVITY 2 Math page 54 85 minutes	**TEACHERS LEARNING THE MATHEMATICS THEY TEACH** Administrators work in small groups to explore the part/whole relationship in fractions. In this context, they consider why teachers must move beyond knowing *how to do* the mathematics procedures to understanding the ideas behind those procedures if they are to effectively help children make sense of mathematics. Through this activity, they get a flavor of what teachers can experience through professional development that incorporates inquiry-based explorations.
ACTIVITY 3 Discussion page 62 45 minutes	**UNDERSTANDING CHILDREN'S MATHEMATICAL THINKING** This activity focuses on what teachers need to learn about listening to children's mathematical thinking. Administrators reflect on their own mathematics learning in the previous activity, and they examine examples of children's confusion about the same topic (fractions as parts of a whole). They discuss what teachers need to know in order to understand children's mathematical thinking, and they consider the characteristics of professional development that could help teachers learn this.
CLOSING page 67 15 minutes	**BRIDGING TO PRACTICE** Participants finish the session with guided reflective writing that helps them link the ideas in Session 2 to their own work as administrators. **HOMEWORK** Administrators will consider further the two components of teacher learning discussed in this session. They will observe three or more teachers in their own schools and write a paragraph about the mathematical understanding of each one. They also read a profile of a teacher who is taking a course to further develop her mathematical understanding (Reading 3).

BIG IDEAS	This session explores
• the need for teachers to understand the mathematics they teach
• the need for teachers to cultivate their ability to listen to students' mathematical thinking and reasoning
• the characteristics of professional development experiences that could help facilitate teacher learning
MATERIALS	☐ Graph paper
☐ Colored pencils or pens
☐ Flip chart and markers
☐ Interlocking cubes
☐ Reading 2 for *Lenses on Learning, Module 2*
☐ Handouts 4–6, pp. 68–72
☐ Bridging to Practice display chart or transparency
PREPARATION	• Review Reading 2, "Learning to Hear Voices: Inventing a New Pedagogy of Teacher Education," to prepare for the discussion of this homework. Referring to your journal as needed, think about the ideas that people in your class put forth in the first session about dimensions of teacher change. Try to anticipate how those people will react to this article. You might be surprised by what some of them say, and you may hear new and different issues come up in their discussion of this reading.
• Prepare and post an agenda for the session, noting the time allotted for each activity.
• Work through the math on Handout 4, *A Cut Above*, to prepare for the work with fractions in Activity 2.
• Activity 3 is somewhat unusual in that it requires you to publicly reflect on your role as facilitator in the previous activity, in which the administrators were working with mathematics. Spend ample time with the agenda and facilitator notes for Activity 3 before you teach this session so that you are conscious of your role during Activity 2.
• Read through Handout 5, *How Children May Think About Fractions*, and think through for yourself the issues that administrators will be discussing in small groups in Activity 3.

OPENING
5 minutes

> **Materials**
> Posted agenda

Introducing the Session

5 minutes

Three areas of teacher learning After participants have assembled, briefly review the three areas of teacher learning that will be discussed today and at the next session of *Module 2:*

- Teachers need to learn for themselves, in a deep and flexible way, the mathematics that they will be teaching.

- They need to learn how to "hear" and make sense of the mathematical ideas that are embedded in children's talk and written work and to decide what instructional moves to make to extend their thinking.

- They need to be able to facilitate discussions in the classroom that further the development of children's mathematical thinking.

These are just some of the new orientations and skills that teachers will need in order to teach in ways that support the development of children's deep understanding of mathematics. Participants need to keep these in mind as they seek to make sense of what the teachers in their schools are grappling with.

ACTIVITY 1
Readings
30 minutes

> **Materials**
> Reading 2
> Handout 3
> Flip chart and markers

One Professional Development Experience

30 minutes | **Whole-group discussion** Invite participants to discuss Reading 2, the article that they read for homework. Remind them of the information that introduced the reading. You might say,

The Heaton and Lampert article is not included in this course as a literal proposal for a model of professional development. Rather, it was assigned to get you thinking about new possibilities for professional development. This is just one way it might play out for a teacher who is working with a committed teaching coach.

You might begin by asking participants to summarize the article. Use the questions on Handout 3 to guide the discussion:

How did Heaton's pedagogical beliefs shift?

What emotional reactions did Heaton have to the things she was learning?

What is it about Lampert and Heaton as individuals that allowed them to establish such a productive relationship?

List participants' ideas on flip-chart paper so that they can get a sense of the range of ideas held in the group.

THE BIG PICTURE
The Heaton and Lampert article was chosen as a reading because it personalizes the ideas of the course with the experience of an individual (and particularly articulate) teacher. It is a good follow-up to the Goldsmith and Schifter article (Reading 1), which addresses the same ideas more abstractly.

TRANSITION | Tell participants that the next activity will be a mathematical exploration that will help them consider the nature of mathematical understandings that teachers need to develop.

Facilitator's Notes
Activity 1

Whole-group discussion

For this discussion, you will need to help the group stay focused on the general professional development issues that the reading illustrates. It is tempting for participants to read this article as a literal proposal for a model of professional development. Once group members get caught on this idea, they may dismiss the model as too intensive and time-consuming for their district.

If participants are having difficulty disconnecting the general points from the specifics of the case, ask them to consider which elements of the Heaton–Lampert collaboration they could imagine actually happening in their schools.

General themes and principles that emerge from this article and could be considered for a school-based collaboration include the following:

- viewing the teaching coach as both a mentor and a colleague
- giving the classroom teacher space to be a learner while respecting that teacher for the classroom experience he or she already has
- recognizing the listening and observation skills a coach needs to bring to a mentoring situation

You might see other generalizations that parallel your own experience.

The following possible responses to the discussion questions are not meant to be definitive, but to give you a sense of the direction the discussion might take. Participants may raise some very different but equally important points. Listen for these ideas and work with them as appropriate, given your judgment of the group's individual and collective understanding of the issues.

◆ **How did Heaton's pedagogical beliefs shift?** Heaton was coming to understand that her role as a mathematics teacher was far more complex than she had assumed. Previously, she had viewed the teaching of mathematics as straightforward: She would give assignments from a textbook and have students work on repetitive problem sets. Now she was realizing that the teaching of mathematics could be more like her teaching of social studies, which emphasized inquiry and encouraged students to reason and ask questions.

◆ **What emotional reactions did Heaton have?** Heaton had to grapple with the feelings of being an inadequate teacher, whereas before she viewed herself as an experienced and even good teacher. As she wrote, "I began to feel dismayed with my own teaching." She also had to deal with the feelings of inadequacy and anxiety as she began to recognize all the mathematics she did not understand and the challenges that lay ahead for her.

◆ **What allowed Lampert and Heaton to establish such a productive relationship?** Both Heaton and Lampert brought to their collaboration a genuine interest in learning. Neither saw herself as an expert with all the answers. Heaton had already experienced a pedagogical style of reflective learning and inquiry through her social studies program and thus was primed to recognize the connections between that experience and the teaching of mathematics. Lampert had developed a mentoring approach in which she was not only a coach or teacher–educator, but also a learner herself. This approach—and her recognition that she had much to learn in working with Heaton—allowed their relationship to develop into a respectful, professional collaboration rather than a more traditional expert–novice relationship.

About the Reading

Reading 2
Learning to Hear Voices: Inventing a New Pedagogy of Teacher Education

This article describes the way schoolteacher Ruth Heaton relearned how to teach mathematics and, in the process, redefined herself as a teacher (and, subsequently, as a teacher–educator). Specifically, she came to a different understanding of what mathematics entailed and what was involved for students to learn mathematics. She learned what her role as a teacher was, both in terms of preparing to teach and being "in the teaching moments" with the students. Heaton also developed new images of collegial relationships, marked specifically by collaborative inquiry, through her work with teacher–educator Magdalene Lampert.

Before her work with Lampert, Heaton had viewed—and taught—mathematics as a body of rules and procedures that students needed to master. With Lampert's help, she came to understand that problem solving and conceptual understanding are central to learning mathematics.

One aspect of the article to highlight is Heaton's and Lampert's choice to ground the sometimes abstract ideas of teacher education in the very real context of Heaton's actual teaching practice. This choice reflects the understanding that teachers need support for learning within the immediacy of teaching and that teachers' knowledge of their craft is, in Lampert's words, "contextual, interactive, and speculative."

ACTIVITY 2
Math
85 minutes

> **Materials**
> Handout 4
> Graph paper
> Colored pencils or pens
> Interlocking cubes
> Flip chart and markers

Teachers Learning the Mathematics They Teach

5 minutes

Introducing the activity In this mathematical investigation, participants will be trying to uncover some of the underlying principles of fractions. The purpose is for them to stretch their own mathematical thinking. Emphasize that the activity is *not* something to do with students but is an experience for adult learners. It will help them expand their knowledge of one area of mathematics while also experiencing what it feels like to learn mathematics in a manner supported by the *Standards*.

Encourage your group to dig into these explorations for their own sake. The follow-up discussions will tease out the implications of this experience for the nature of mathematical understandings teachers need to develop and the nature of professional development that will provide that teacher learning.

THE BIG PICTURE
Through this exploration, administrators experience what it means for teachers to move beyond merely knowing *how to do mathematics procedurally* to *understanding the mathematical ideas behind the procedures*. They also consider the wider implications for the professional development of teachers.

35 minutes

Doing the math in small groups Distribute Handout 4, *A Cut Above*. Divide participants into small groups of two or three. They are to spend about 30 minutes on the two problems, using drawings or cubes to work out their answers. Emphasize that they must use these visual or physical methods instead of straight numerical reasoning, although they can use numbers to help represent their findings and to show how they got there.

As the participants work through the mathematics in small groups, be prepared to step back and observe yourself facilitating. Reflect on the things you notice and the decisions you make as you listen to participants and try to make sense of what they understand. You will describe this experience in Activity 3 to help administrators appreciate a teacher's role in the classroom. See pages 62 and 64 for questions you might consider. Take notes as needed for your discussion during Activity 3.

The two problems are as follows:

1. *At Piece o' Pizza, the pieces in small, medium, and large pizzas are all the same size. A small has 4 pieces, a medium has 6 pieces, and a large has 8 pieces. You order a medium garlic-artichoke pizza and a large spinach-pineapple pizza. You eat $\frac{2}{6}$ of the garlic-artichoke and $\frac{3}{8}$ of the spinach-pineapple. How much of the pizza did you eat?*

SMALL   **MEDIUM**   **LARGE**

2. *At Creative Cuts Pizza, all the pizzas are the same size, but they are called Small, Medium, or Large depending on the size of their pieces. Thus, there are 4 large pieces in a large pizza, 6 medium pieces in a medium pizza, and 8 small pieces in a small pizza. You order a medium garlic-artichoke pizza and a small spinach-pineapple pizza. As before, you eat $\frac{2}{6}$ of the garlic-artichoke and $\frac{3}{8}$ of the spinach-pineapple. How much of the pizza did you eat this time?*

LARGE   **MEDIUM**   **SMALL**

THE BIG PICTURE
Rather than focusing on algorithms for performing operations on fractions, a goal of this exercise is for administrators to understand some of the conceptual underpinnings of work with fractions.

Circulate among groups to get a sense of how participants are approaching the problems, what discoveries they are making, and what difficulties they are encountering. This will help you shape the whole-group discussion.

Note that the questions are deliberately ambiguous. If participants ask you to clarify, explain that this is part of the investigation and encourage each small group to make their own interpretation.

Remember that later you will be describing your own process of listening and reflecting during this activity.

30 minutes

Whole-group discussion of the problems Call participants together and ask them to report the answers their group found for problem 1. There will likely be a variety of answers. Without commenting, record each answer on a flip chart visible to all.

Focus on several answers that were the same and ask about the groups' strategies. You might say,

> *Are there different ways to get the same answer? Let's see what strategies were used by those who arrived at the same answer.*

Next, look at answers that were different. You might ask,

> *Could more than one numerical answer be correct? If yes, why? If not, why not?*

Invite participants who had different answers to explain their thinking. Ask the group to consider why, for this problem, other answers might also be correct. Help them ask questions that encourage others to *explain* rather than *defend* their thinking.

Ask participants to describe what puzzled them about this problem, what blind alleys they went down, what assumptions they had made when they went down those blind alleys, and what made them change their minds. Work with them to uncover why their original assumptions might have made sense in another context, and why they didn't work in this one.

THE BIG PICTURE
This activity takes learners far beyond what any set of isolated fraction problems on a page could reveal. It underscores for administrators the importance of situating work with complex mathematical ideas in contexts that have meaning for children. It also illustrates the importance of pausing in mathematical explorations to ask the question, "Does this answer make sense?" Those who apply mathematical procedures blindly often lose the insight that their common sense could offer them.

Address the second problem in a similar manner, first collecting the range of answers, then asking about the strategies used by those who arrived at the same answer, and finally asking about approaches used by those who arrived at different answers. Again, encourage participants to share the process they followed, describing their "false starts" and why they think those didn't work.

As participants share and discuss their findings, they should begin to realize that the variety of answers results from the ambiguous nature of the questions.

Encourage participants to reflect on the process by which they began to sort out the mathematics of fractions. You might ask,

At what points in the process did you stop and ask yourself, or another member of your group, whether the answer made sense?

THE BIG PICTURE
A main goal of this activity is to highlight the nature of mathematical understanding that teachers need. The final part of this discussion ties the previous mathematical exploration to this central issue. It is important to leave enough time to address the role of mathematics learning in professional development for teachers.

15 minutes	**Teacher learning in mathematics** Finally, shift the conversation to the role of mathematics learning in teacher professional development. You might ask,
	If teachers did not understand these mathematical ideas about fractions, how might that interfere with their teaching of fractions?
	What type of professional development would help teachers develop understandings like these? What would it need to incorporate?
	Would it help teachers to go through the activity you just experienced?
TRANSITION	At the conclusion of the discussion, tell participants that the next activity will help them reflect on the complex process of understanding someone else's mathematical thinking, another key element of teaching practice for standards-based mathematics.

FACILITATOR'S NOTES
ACTIVITY 2

Doing the math in small groups

The mathematics involved in the two pizza problems is discussed in detail in About the Mathematics, page 60. Be sure to read this before presenting the activity.

As you circulate among the small groups, you may hear administrators wonder how doing a mathematics exercise like this connects to their own administrative practice. You might explain that while some of them may learn something about operations with fractions, this activity is not intended to improve their quantitative reasoning skills. Rather, the point is to sensitize administrators to the complexity of elementary mathematics and to the type of mathematical explorations teachers need to facilitate with their students.

You may see different reactions to this work. One reaction to be prepared for is confusion. In Session 1, administrators may have *discussed* the role of confusion, but in this activity, they are likely to experience it for themselves. You will want to be prepared for the reactions; some administrators may become engaged and excited with the mathematics, being drawn into figuring out how the fractions work, while others may become frustrated or even angry.

In particular, monitor the reactions and participation of any mathematics specialists in the group. This work may be confusing even for them, and they could feel especially vulnerable in the presence of colleagues. On the other hand, math specialists who are particularly comfortable with the mathematics can be looked to as a resource. They can help with guiding people through the mathematical exploration. However, watch for administrators with mathematics backgrounds who start to play the role of expert. This stance could be very intimidating to other participants and might limit their mathematical exploration. As the facilitator, you will have to monitor the social dynamics.

The activity is structured to be deliberately ambiguous, and some administrators may wonder if children could handle the ambiguity. You can point out that this was designed as an adult activity, although the level of ambiguity would be appropriate for upper elementary grade levels. For students at this level, ambiguity can foster mathematical discourse and lead to an appreciation for the place of argumentation and proof in mathematics.

Whole-group discussion of the problems

You will follow the same approach to both problems.

◆ **Are there different ways to get the same answer?** Participants will likely approach these problems and represent their solutions in different ways, some numerically and others with a drawing or cubes. By sharing and following each other's strategies, administrators enrich their understanding of the mathematics at play and also experience how a mathematics classroom can be an intellectual community.

◆ **Could more than one numerical answer be correct?** Because of the ambiguity in the problem, several different answers can be correct, as discussed in About the Mathematics. Someone might ask, "So, what is the *right* answer?" Help participants arrive at an understanding that the correct answer depends on the particular whole that is being considered.

You might also point out that while several different answers are possible, that doesn't mean that just *any* answer can be made to fit. Administrators sometimes get caught up in an all-or-nothing mentality, thinking that if there isn't just one right answer, then any answer is okay, or the answer doesn't matter. Point out that an answer has to be mathematically consistent within the parameters of the problem but that sometimes more than one answer can be mathematically justified.

Some participants may have arrived at the same answer ($\frac{17}{24}$) for both problems if they applied the "find-a-common-denominator" algorithm, even though they were working with different wholes. Even if no one did this, you might raise this possibility and ask why the same answer for both problems would be problematic.

Encouraging participants to reflect on their misconceptions helps them arrive at a fuller understanding of the mathematics of fractions. It also helps others reconsider incorrect strategies that they have not yet recognized. As one *Lenses on Learning* facilitator reflected,

> *In the few mathematics explorations that we're able to fit in, we want [participants] to get a taste of the complexity of the ideas behind the facts and procedures of elementary math, to develop some perspective on why kids struggle with them, and to come away with some new insights or new ways to think about them.*

♦ **At what points did you stop and ask yourself whether the answer made sense?** This question underscores the importance of using common sense in solving mathematical problems and the dangers of blindly applying mathematical procedures. If some administrators admit to not considering this question, ask why they didn't think to do this, and why it is always an important thing to do.

Teacher learning in mathematics

In this final portion of the discussion, help administrators make the link between the mathematical work they did and what their teachers might need for professional development in mathematics.

♦ **If teachers did not understand these mathematical ideas about fractions, how might that interfere with their teaching?** Teachers need to have a broad view of the mathematical ideas about fractions that students need to develop through the elementary years. Without such a view to guide their teaching, teachers might either skip over ideas that students ought to address in depth, or pursue ideas that are inappropriate to the age and abilities of the students. Equipped with a long-term mathematical agenda, as well as an understanding of the misconceptions some students are likely to have and the ideas they are apt to grasp easily, teachers are in a better position to help a range of students build robust understandings of the mathematical topics covered in any given year.

♦ **What type of professional development would help teachers develop understandings like these?** This final question brings the discussion back to some of the big ideas of this session. A suitable professional development plan would help teachers deepen their understanding of the mathematics; help them become more knowledgeable about the range of children's mathematical thinking they are likely to encounter in their classrooms; and help teachers further develop their abilities to work with students' mathematical ideas.

At this juncture (and at other points in this module), participants may express a concern that the teachers in their schools could never develop the expertise to teach mathematics effectively. They may wonder whether having math specialists to teach all the math classes, especially at upper elementary levels, wouldn't be a better way to work with the limitations of their current staff. Clearly, there are no universal answers to such dilemmas, but there will be benefits and trade-offs to any solution. Your role is to help participants make sense of the ideas and perspectives that are at play here, rather than attaching themselves to any quick solution.

A separate but related issue concerns the role of technology in the teaching of mathematics. The appropriate use of computers and calculators is among the issues that teachers are being asked to rethink. Administrators need to be acquainted with professional development opportunities for teachers that combine learning how to master new tools with reflections on how their use can complement, and not detract from, the conceptual work students must do. This issue may come up again in Session 4, when participants read about and discuss electronic professional networks.

About the Mathematics

Fractions, although conceptually complex, are often taught procedurally or algorithmically. That is, we have all been taught rules for adding, subtracting, multiplying, and dividing fractions, but we don't often have the chance to think about what types of numbers they represent, how they behave, and what functions they serve.

In part, fractions are complicated because even simple fractions are used in a variety of situations. They may describe part of the area of a shape (three of eight parts of a rectangle), part of a group of things (13 children out of a class of 25), part of a length (three-quarters of the distance from one end of a field to the other), or a rate (11 train passes for the price of 10). While these contexts offer good models for visualizing fractions, their range can be confusing.

This activity uses an *area* model to emphasize several central concepts about fractions.[1]

Fractional parts of a whole are equal parts. In problem 2, the pieces of pizza are not equal and thus cannot be compared or added until we recast them in twenty-fourths of each pizza (or forty-eighths of the two pizzas combined).

Equal parts of shapes are not necessarily congruent. Again in problem 2, when we cut the pieces further to get equal-size fractions, pieces that represent $\frac{1}{24}$ of a pizza will have the same area, even though their shapes may be different.

Two fractions (relative to the same unit whole) can be added or compared. In both problems, once the size of the pizza pieces is consistent and we are clear about what *whole* is being referred to, we can add or compare the fractions of the whole they represent.

In this investigation, participants discover the importance of consistently referring to the same whole when adding fractions. The directions in problems 1 and 2 are deliberately ambiguous, and participants are left to define for themselves what constitutes the whole they are working with. In the question "How much of the pizza did you eat?" we don't know if "the pizza" refers to the two medium and the large pizzas that were ordered, to those two pizzas separately, or to all three sizes of pizza that are pictured (small, medium, and large).

Problem 1: Piece o' Pizza

The first problem involves equal-sized *pieces* of pizza, but different-sized whole pizzas. Here, you might designate as "one whole" either all the pizza that is pictured here ($\frac{18}{18}$); all the pizza you *bought* ($\frac{14}{14}$); or each pizza you ordered—the medium ($\frac{6}{6}$) and the large ($\frac{8}{8}$). Regardless of your choice, you can add the individual pieces together and describe them in reference to the whole you have chosen: you ate $\frac{5}{18}$ of all the pizza pictured; you ate $\frac{5}{14}$ of all the pizza ordered; or you ate $\frac{2}{6}$ of a medium and $\frac{3}{8}$ of a large.

Participants who are used to the "find-a-common-denominator-and-add-them-up" algorithm might do something along these lines:

$\frac{0}{4} + \frac{2}{6} + \frac{3}{8}$

Uh-oh, the denominators are all different. I need to find the common denominator:

$\frac{0}{24} + \frac{8}{24} + \frac{9}{24} = \frac{17}{24}$

---

1. These ideas are adapted from Cornelia Tierney, Mark Ogonowski, Andee Rubin, and Susan Jo Russell, *Different Shapes, Equal Pieces*, a grade 4 unit of *Investigations in Number, Data, and Space* (Glenview, IL: Scott Foresman, 1995).

In this case, applying the standard algorithm for adding fractions gives a total that is greater than $\frac{1}{2}$, even though only 5 out of 14 pieces ordered (or 5 out of 18 pieces altogether) were eaten.

If some administrators do the problem this way and no one points out the discrepancy, point it out yourself. Ask why the standard algorithm is not working here. For further guidance, you might ask, "Which *wholes* do you think you're talking about with each of your equations?"

Recognizing the whole is a critical part of understanding fractions. While fractions make it possible to *compare* wholes of different sizes, we need to be consistent about the whole when trying to add or subtract fractions.

Problem 2: Creative Cuts Pizza

The second problem involves adding different-sized pieces of same-sized pizzas. One of the key concepts about fractions is that fractional parts of a whole need to be equal, though not necessarily congruent. When we find a common denominator by redrawing the lines to divide both the medium and the large pizzas into 24 equal parts, we might end up with differently shaped pieces (some squat and rectangular, others tall and skinny), yet these are still equal-sized pieces: $\frac{1}{24}$ of a same-size whole. As was the case in problem 1, we need to keep in mind which whole we are working with before we add any fractions. Thus, depending on which whole we consider, the answer might be $\frac{17}{24}$ of a single pizza, $\frac{17}{48}$ of two pizzas combined, or $\frac{17}{72}$ of all three pizzas combined.

Like percentages and averages, fractions are often used for comparison. This is a complex idea, distinct from just comparing actual amounts (such as how many more cookies Jana has than Jim), for which we use an operation like subtraction to find a difference. The whole can be any size, as long as we are comparing fractional parts of the whole *with fractional pieces that are the same size*. That is, we can compare $\frac{2}{3}$ and $\frac{1}{2}$ of the same whole if we express both as sixths. Activities like this one that explore the underlying qualities of fractions lay a foundation for understanding how such comparisons work.

ACTIVITY 3
Discussion
45 minutes

Materials
Handout 5

Understanding Children's Mathematical Thinking

10 minutes

Facilitator's reflections on listening Give a quick overview of the activity, explaining that participants will have a chance to discover how challenging it can be to attend closely to someone else's mathematical thinking for evidence of real understanding. To start, you will reflect on your experience with them in the fraction activity. By hearing about the listening *you* did in order to understand *their* thinking, administrators may better understand the listening that teachers need to do in order to understand children's thinking.

THE BIG PICTURE
Listening to children's mathematical thinking is a new skill for many teachers—and administrators as well. Those who have been through *Lenses on Learning, Module 1* were introduced to the skill of listening to a student's thinking when they watched a video clip of a clinical interview with a fifth-grade child. This activity gives administrators a broader, experiential view of what it means to listen to someone else's mathematical thinking and to consider the instructional moves that might help extend a child's thinking.

Share your own process of listening during the fractions activity, explaining how you made sense of what people in small groups were understanding. Your reflections might include the following ideas:

What were the mathematical ideas you found most intriguing?

What was most challenging for you to do or to understand as you moved among the groups?

How did the listening help you to facilitate the whole-group discussion and make decisions about the mathematical points to emphasize?

How did you take into account both the mathematical concepts and the heterogeneity of your group in facilitating the whole-group discussion?

Emphasize that an understanding of the mathematics was central to your ability to appreciate and understand participants' ideas and to decide how to work with them.

20 minutes	**Small-group work: Examining children's thinking** Ask participants to gather in the same groups of two or three they had for the previous activity. Any pairs should join to form a group of four. Distribute Handout 5, *How Children May Think About Fractions*. Groups should discuss the three questions on this handout: *What are the children in each example confused about?* *How does the confusion illustrated in each example relate to any confusion that you had when doing the pizza problems?* *How might teachers' own struggles to make sense of mathematics help them understand what their students are struggling with?* Circulate to listen to small-group discussions. Try to identify the mathematical ideas they are talking about—for example, the fact that any particular fraction occurs in relation to a particular whole, and that if children don't keep this in mind, they can become confused. In the subsequent discussion, you may want the whole group to consider certain ideas, or you may want to bring up a mathematical idea that was *not* discussed, asking, "Did anyone think about X?"
15 minutes	**Whole-group discussion** Bring the groups together for a general discussion about listening to children's mathematical thinking and the implications for teachers. You might ask, *What do you think teachers would need to know about fractions to listen to students' mathematical thinking in the way you just did by analyzing the thinking in evidence on Handout 5? How might they learn this?* Administrators should mention ideas such as understanding an area model for fractions and the problems that students have in understanding an area model. This discussion should also touch on the type of professional development that would promote careful, discerning listening.
TRANSITION	In the next session, the group will take a closer look at listening and its role in facilitating mathematical discourse. In the last two sessions, they will look at the types of professional development that can help teachers in this area.

FACILITATOR'S NOTES
ACTIVITY 3

Facilitator's reflections on listening

The first segment of this activity, your *reflection* on your role as facilitator in the previous activity, may feel uncomfortable to you. However, it can be valuable for administrators because it helps them understand and appreciate what a reflective stance entails.

As you share your reflections on listening, keep in mind any parallels between the work you did to make sense of the administrators' thinking and the work teachers do to make sense of students' thinking. Following are some points you might share.

- **Intriguing mathematical ideas** When you listened to the groups, did you hear strategies that were surprising to you, or strategies that actually helped you to understand the problem better? For example, did someone draw a diagram that seemed to capture the essence of the problem in a way that you hadn't visualized before? What was mathematically interesting about it?

- **Challenges** What did you find challenging about listening to the different groups? For example, was a particular strategy more difficult to follow than you expected? What might be challenging for teachers in such a situation? Was it challenging to keep track of different strategies? If so, what did you do to manage this? Did you select a few strategies on which to focus? Did you choose to focus on a general conceptual issue, such as how participants visualized what the whole was?

- **Facilitating discourse** How did you use what you heard in the small groups to guide the whole-group discussion? For example, how did you incorporate the specific strategies you heard? Did you choose to bring up (or not bring up) an issue because it didn't come up in the small groups?

- **Heterogeneity** Were there any specific issues regarding the heterogeneity of the group that shaped your choices for the whole-group discussion? For example, did you choose to not pursue a particular strategy because it seemed too mathematically advanced for some people in the group? Were you careful not to highlight someone's mathematical struggles?

Small-group work: Examining children's thinking

In discussing Handout 5, administrators have the opportunity, once again, to think about children's mathematical knowledge.

- **What are the children in each example confused about?** The children whose mathematical thinking is illustrated on Handout 5 are struggling to get a firm conceptual grip on ideas related to fractions. In particular, they are wrestling with the idea that a fraction is *relative*. It is a quantity expressed *in relation to* a particular whole; for example, $\frac{1}{3}$ is a certain quantity that is being described in terms of its relation to a particular whole. The three examples illustrate different ways in which children can experience this as a slippery idea.

Example 1 The first child, Kim, has lost track of what is the whole and what are the parts. She has forgotten that her "class strip" already represented the whole class, and so when figuring out what $\frac{1}{2}$ is, she does not fold her class strip in half, but adds a second class strip and declares that the original strip is $\frac{1}{2}$ of the new whole. She reasons that if she had three strips, the original strip would be $\frac{1}{3}$ of the new whole.

Example 2 The second example, finding a familiar fraction to express 5 out of 20, illustrates how the notion of the relationship between parts and wholes can come in and out of focus for learners. Karen and David at first say that $\frac{1}{3}$ of the class is ten years old. Karen explains that she "counted how many fives are in the 15 and divided it." That is, she counted and discovered that there were three fives in 15, and converted the three into $\frac{1}{3}$. However, she has chosen the wrong whole, using 15 instead of 20. David then

64 ♦ Session 2, *What Do Teachers Need to Learn?*

gets the right answer, saying that it is $\frac{1}{4}$. David understands that the whole is 20, which is made up of 15 nine-year-olds and 5 ten-year-olds. He has taken the 15 that Karen used and "included all the ten-year-olds," adding 5 to get 20. They then launch into a new question, asked by Jesse, "What fraction of the class is nine years old?" David, who understood parts and wholes earlier, now thinks that the whole is 15—three groups of 5. In the end, David and Karen together work out that there are four groups of 5 (three groups of nine-year-olds and one group of ten-year-olds), so that $\frac{3}{4}$ of the group is nine years old and $\frac{1}{4}$ of the group is ten years old.

Example 3 Tuong counts off by four, the numerator of the fraction $\frac{4}{22}$, and concludes that there are four parts (each comprised of four sections) to the whole strip rather than five parts. He seems to be thinking that $\frac{1}{4}$ has to have four sections in it—that a part is the number of sections per fold rather than the number of folds per strip. In short, he is confused over what $\frac{1}{4}$ means; he is confusing the number of sections per part with the number of parts per strip.

♦ **How does the confusion illustrated in each example relate to any confusion participants had?** These examples raise issues similar to those that participants encountered when working on the pizza problem—questions about what is the part and what is the whole. Participants should easily find parallels.

♦ **How might teachers' own struggles to make sense of mathematics help them understand what their students are struggling with?** Through the pairing of these two activities (doing the pizza problems and reading children's ideas about fractions), administrators should see that their own mathematical struggles (and, by extension, teachers' struggles) can help them understand what children find difficult about mathematical ideas. They should be able to say from their own experience what teachers need to know in order to listen sensitively to children's mathematical thinking.

Whole-group discussion
Point out that the attentiveness with which you listened to administrators' mathematical thinking during the pizza activity, and the attentiveness they brought to the transcripts of children's thinking about fractions, are both similar to the listening teachers must do. Such listening is a multifaceted process. While critics sometimes characterize standards-based mathematics teaching as employing a passive "anything goes" stance, in fact this kind of listening brings together a number of interrelated processes that teachers then use to determine appropriate instructional moves. When teachers are truly attending to what children are saying, they are

- assessing the mathematical validity of children's reasoning
- situating the children's thinking and problem solving in a larger mathematical frame of reference
- identifying one or more particular mathematics concepts, beyond the immediate problems at hand, that children are grappling with[2]

♦ **What do teachers need to know in order to listen to students' mathematical thinking?** Administrators may mention the mathematical ideas they just worked on: understanding an area model for fractions, and the ability to identify what is the whole and what are the parts. They may also mention more general skills, such as being able to assess the mathematical validity of children's reasoning and situating it in a larger mathematical frame of reference.

---

2. These characteristics of teacher listening were suggested by Deborah Schifter, personal communication.

♦ **How might teachers learn this?** In this activity, administrators have experienced a kind of professional dialogue about mathematical thinking that may be new to them. Some may feel that such discerning and intent listening is beyond what would be possible in their district. Others may feel that their district is so politically charged, they cannot imagine teachers being willing to work together, or with them, to cultivate such listening skills.

Help participants begin to generate ideas about the kinds of professional development experiences that would support teachers' development of these skills. Here is another opportunity to return to some of the big ideas in this module: that teachers need opportunities to deepen their understanding of the mathematics, to advance their capacities to understand the range of thinking of the students they teach, and to build on students' mathematical ideas. Encourage participants to relate teachers' needs to their own experiences in this module—what they feel can be gained from doing the mathematics and reflecting together about it, and what can be learned from looking at examples of students' thinking.

CLOSING
15 minutes

Bridging to Practice

Remind participants that the Bridging to Practice exercise enables them to focus on the ideas in the course that seem particularly pertinent to the school community in which they work. This reflective writing gives them a chance to think about particular ideas they could pursue in their own jobs.

Invite participants to respond to the three standard questions:

1. Pick an idea that came up today that you found particularly interesting. What is your current thinking about this idea?

2. Where is your school now with regard to this idea?

3. What are one or two things that you will go back and pursue, to move yourself and/or your school along with this idea?

Homework

Distribute Handout 6. With this homework, administrators further consider the two elements of teacher learning that were discussed in this session: understanding the mathematics they teach and making sense of the variety of children's mathematical thinking. Participants are asked to observe teachers in their own schools who are working on changing their teaching, to see how these skill areas play out in practice. They are to complete a written report on their observations.

A supplemental reading, "Becoming a Mathematical Thinker: Linda Sarage" (Reading 3), offers a powerful, grounded image of a teacher engaged in the learning process.

A Cut Above

Use drawings or cubes to solve these two problems.

1. At Piece o' Pizza, the pieces in small, medium, and large pizzas are all the same size. A small has 4 pieces, a medium has 6 pieces, and a large has 8 pieces. You order a medium garlic-artichoke pizza and a large spinach-pineapple pizza. You eat $\frac{2}{6}$ of the garlic-artichoke and $\frac{3}{8}$ of the spinach-pineapple. How much of the pizza did you eat?

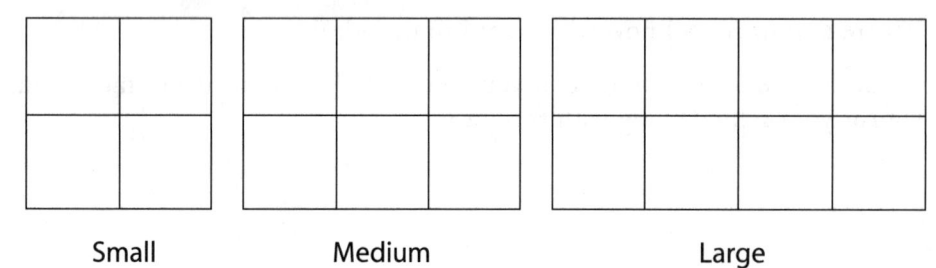

Small   Medium   Large

2. At Creative Cuts Pizza, all the pizzas are the same size, but they are called Small, Medium, or Large, depending on the size of their pieces. Thus, there are 4 large pieces in a large pizza, 6 medium pieces in a medium pizza, and 8 small pieces in a small pizza. You order a medium garlic-artichoke pizza and a small spinach-pineapple pizza. As before, you eat $\frac{2}{6}$ of the garlic-artichoke and $\frac{3}{8}$ of the spinach-pineapple. How much of the pizza did you eat this time?

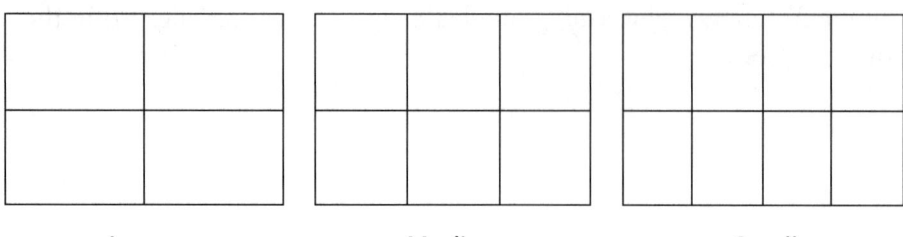

Large   Medium   Small

How Children May Think About Fractions

The three excerpts from children's conversations on the next sheet come from a fourth-grade unit on fractions.* The students have been exploring fractions by thinking about parts of their own class with problems like this one:

> There are 14 people in our class who are wearing red. There are 26 people in our class altogether. What fraction of the class is wearing red?

The students have made "class strips," which are strips of adding machine paper with lines drawn to create sections. Each section represents one student in the class, so the entire strip represents the whole class. Students then fold the class strips into fractional parts: halves, quarters, or thirds. This enables them to describe fractions such as $\frac{14}{26}$ or $\frac{12}{22}$ in more familiar terms; for example, "a little more than $\frac{1}{2}$."

The three examples of student thinking on this handout illustrate the complexities of fractions. Note that these examples are from two different classes, so the total number of students in Example 2 is different from the total number in Example 3.

In your small group, discuss the following questions:

1. What are the children in each example confused about?

2. How does the confusion illustrated in each example relate to any confusion that *you* had when doing the pizza problems?

3. How might teachers' own struggles to make sense of mathematics help them understand what their students are struggling with?

---

* Adapted from Mary Berle-Carmen, Karen Economopolous, Andee Rubin, and Susan Jo Russell, *Three out of Four Like Spaghetti,* a grade 4 unit of *Investigations in Number, Data, and Space* (Glenview, IL: Scott Foresman, 1995).

EXAMPLE 1

Kim: I think it's $\frac{1}{2}$. If you put two class strips together, then this one [she picks up the one she has marked off the data on] is $\frac{1}{2}$. If you have three strips, then this one is $\frac{1}{3}$, so it's $\frac{1}{2}$ because it's not $\frac{1}{3}$. If you had four strips, it would be $\frac{1}{4}$.

EXAMPLE 2

In this conversation, another group of students move in and out of clarity about what the whole is and what fractional part they are considering. There are 20 students in their class. They are trying to find the "familiar fraction" for 5 out of 20, which is the number of students who are ten years old. (The other 15 are nine years old.)

Karen: I think it's $\frac{1}{3}$.

David: Yeah. One-third of our class is ten years old.

Teacher: How did you figure that out?

Karen: I did 5 times what is equal to 15. No, that's not what I did. I think I counted how many fives are in the 15 and divided it. So I got three fives, and one part is $\frac{1}{3}$.

David: It's actually $\frac{1}{4}$ because you have to include all the ten-year-olds. There are 20.

Jesse: What fraction of the class is nine years old?

David: Three-thirds.

Karen: No, $\frac{2}{3}$. Wait, it is $\frac{1}{4}$ isn't it?

David: Yes, because 3 groups of 5 and then 1 more 5.

Karen: So $\frac{3}{4}$ of the group is nine years old and $\frac{1}{4}$ is ten years old.

EXAMPLE 3

There are 22 students in this class. Tuong is trying to figure out a familiar fraction for $\frac{4}{22}$. He writes the numbers 1 through 22 in the 22 sections of his strip. Then he counts off four sections and folds the strip over and over, so there is a fold after every four sections. He ends up with a strip that is folded into fifths with two sections left over. He explains: "I was trying to get fourths and since there were four sections I folded on every fourth one." He has made fifths but called them fourths.

Homework for Session 2

Written Assignment: Profiling the Teachers in Our Schools

In today's session we talked about two areas in which teachers need to develop new skills: understanding the mathematics they teach and making sense of the different mathematical ideas that children have.

Your homework is to observe three or more teachers in your school, choosing some you consider to be weak in one or both of these areas and some you perceive to be strong. (Some teachers may be strong in one area and need a great deal of work in another.) Try to choose a group of teachers who, taken together, exhibit a range of strengths and weaknesses in the two areas.

You should get this information by observing teachers in their classrooms, by paying attention to hallway and other impromptu conversations, and by listening closely to what people say in a faculty meeting that focuses on teaching.

Be clear about the criteria you are using to make your judgment, but think of your profile as a set of conjectures, based on the information you currently have, which can be changed later as you acquire additional information. For this assignment, you are writing a profile of the teachers based simply on what you know now.

Write a paragraph about each teacher. Tie your descriptions to specific conversations or observations. Also address the following questions:

- What are the benefits to a teacher who is comfortable with the mathematics content?
- What are the benefits to a teacher who is comfortable listening to children's thinking?
- Are there connections between these areas?

Reading Assignment: A Teacher's Learning

As a supplemental assignment, please consider the following:

READING 3 "Becoming a Mathematical Thinker: Linda Sarage" by Deborah Schifter and Catherine Twomey Fosnot

This excerpt from the book *Reconstructing Mathematics Education* describes a powerful professional development program, mathematical explorations of the division of fractions, and the cognitive and affective aspects of an individual teacher's learning.

If you are pressed for time, you might just skim Reading 3. Please spend the bulk of your time on your observations and written teacher profiles because that work will be a central part of an activity in the next session.

SESSION 3

What Makes for Meaningful Professional Development?

Thus far in this module, administrators have explored two important areas of learning for teachers: deepening their understanding of mathematics and learning to listen to students' mathematical thinking. A third important area of teacher learning, facilitating mathematical discourse in the classroom, is explored in this session.

Facilitating mathematical discourse requires active listening—being able to follow students' thinking and to make judgments about the validity of their ideas. Only then can teachers navigate through the subtle, moment-by-moment decisions that are inherent in teaching and facilitating a discussion: which of the children's ideas to build on, which questions might be intellectually generative, what level of language complexity best serves the whole group of children, and how to interpret and work with students' confusion or their silence.

By this point in *Lenses on Learning,* administrators are familiar with new ideas about learning and teaching and with the complexity of what teachers need to learn. Most of the teachers they have observed in this course—on videotapes or in excerpts from their writing—have been seriously interested in the new ideas and have begun the process of changing their instructional practice. Even though most of these teachers are at an early stage of their learning, administrators in your group may feel that not many of their teachers are as far along as those whose thoughts and practice they have encountered in this course. They may wonder what it would take to help move their teachers in this direction and what their own role might be. This session begins to address those concerns.

Overview for Session 3

OPENING page 76 5 minutes	**GETTING STARTED** Begin with announcements and a brief preview of the session agenda.
ACTIVITY 1 Video page 77 50 minutes	**FACILITATING DISCOURSE** Administrators view a video clip of a first-grade teacher facilitating a whole-group discussion. Using this as a jumping-off point, they consider what teachers need to learn to make mathematics discourse come alive in their classroom and to support the development of students' mathematical ideas.
ACTIVITY 2 Homework Discussion page 83 55 minutes	**THE TEACHERS IN OUR SCHOOLS** Participants look at the homework they prepared, which involved observing teachers in their schools. They now extend each profile to reflect the teacher's capacity to facilitate classroom discourse that extends and deepens mathematical understanding. They develop and share visual representations of their findings on strengths and weaknesses in their school. This activity helps administrators move beyond "one-size-fits-all" professional development approaches by making explicit the various areas in which different teachers need to work.
ACTIVITY 3 Video page 87 55 minutes	**LEARNING MATHEMATICS TOGETHER** The video in this activity shows teachers in the *Talking Mathematics* seminar who are exploring and "talking mathematics" themselves. This serves as a starting point for identifying characteristics of meaningful professional development.
CLOSING page 93 15 minutes	**BRIDGING TO PRACTICE** Participants finish the session with guided reflective writing that helps them link the ideas in Session 3 to their own work as administrators. **HOMEWORK** The assignment is a reading that explores the value of collegial forms of professional growth for teachers, in contrast to the "training" model of professional development.

BIG IDEAS	This session explores • the role of classroom discourse in mathematical learning • the varying professional development needs of teachers within any school or district • characteristics of professional development that fosters learning to facilitate discourse
MATERIALS	☐ Flip chart and markers ☐ Highlighters and stick-on notes ☐ Video clips, *Today's Number* and *Talking Mathematics: A Teacher Seminar* ☐ Handouts 7–11, pp. 94–98 ☐ Bridging to Practice display chart or transparency
PREPARATION	• Prepare and post an agenda for the session, noting the times allotted for each activity. • Review your journal notes from the previous session. If you are collecting administrators' reflective writings, review those as well. Make note of any professional development issues administrators have raised that pertain to the ideas in this course. • Preview the two video clips and think about the video discussion questions on Handouts 7 and 10. • For Activity 2, prepare an example of a bar graph that represents the profile of an individual teacher in three areas of learning (mathematics knowledge, capacity to listen to children's thinking, and ease with facilitating discourse).

OPENING
5 minutes

Materials
Posted agenda

Getting Started

5 minutes | **The video experience** As the group assembles, begin with announcements and a preview of the session agenda. You might mention that this session is unusual in that two different video clips will be shown. Explain that these videos offer images of real teachers as they seek to reconceptualize their teaching practices and experience new forms of professional development.

ACTIVITY 1
Video
50 minutes

> **Materials**
> Video clip, *Today's Number*
> Handout 7
> Flip chart and marker

Facilitating Discourse

5 minutes | **Introducing the video clip** Explain that in this video clip, participants are to consider the role of discourse in helping children develop their mathematical thinking and to look for teacher qualities and actions that facilitate this type of classroom talk. Provide the following information about *Today's Number:*

> *The tape was shot in Ms. Scott's first-grade class in the spring. Ms. Scott is in her first year of teaching.*
>
> *The eight-minute clip is an excerpt of a whole-class discussion that took place after the children had worked individually on the problem, coming up with a variety of ways to make the number 16 (corresponding to the day of the month).*
>
> *For several students, English is not their primary language.*

Take a minute to remind participants about bringing a constructive orientation to viewing a classroom episode on video. Encourage them to resist the temptation to make broad judgments about the teacher. Review the principles set forth in Setting a Climate for Viewing Videotapes of Teachers and Students at Work (p. 15), in particular:

> *The central purpose is not to critique the teaching. From the perspective of an outside observer, there are always other moves that a teacher might have made. When we see only a short excerpt of a lesson, we do not know enough to make valid judgments about whether the teacher's decisions were good ones. Rather, the purpose of viewing the video clip is to stimulate discussion of important ideas about mathematics learning and teaching.*

THE BIG PICTURE
The intent of this activity is to point out the teacher's central role in helping children invent their own strategies for doing arithmetic, solve problems and describe their work to others, and hear their classmates' strategies. To make this happen, the teacher must foster a classroom culture in which children can comfortably talk about their ideas.

15 minutes | **First viewing and discussion** After viewing the clip once, invite participants to share briefly with the group what they saw and heard. Be attuned to different and possibly conflicting observations. Ask for evidence from the video for their statements. For example,

> *What did Ms. Scott do that makes you say that? What did you see or hear that leads you to think that?*

Explain that this is not a time to try changing anyone's point of view, but rather to make statements and raise questions. Before showing the video a second time, summarize the issues so that everyone can be open to further insights into the learning that is made possible by the mathematical discourse in Ms. Scott's classroom.

THE BIG PICTURE
In a second viewing of the video, administrators are better able to look beyond the specifics of the activity to see how Ms. Scott works to bridge the range of thinking in the class, building scaffolding so that students whose strategies are more basic are able to share in the more complex mathematical thinking initiated by some of their peers.

30 minutes | **Second viewing and discussion** Distribute Handout 7, *Discussing the* Today's Number *Video*. Suggest that participants take notes on these questions for reference in the subsequent discussion.

1. *What are some of the strategies that Ms. Scott uses to engage a wide range of students in her class?*

2. *How does Ms. Scott work with the strategies of students who are at different points in the development of their understanding of mathematical patterns and relationships?*

3. *What are some questions Ms. Scott asks, statements she makes, or other actions that she takes that guide the discussion in mathematically important ways?*

After the second showing, give participants a minute or two to complete their notes. Then invite them to share their reflections, using the questions on the handout to guide the discussion. As participants mention the things a teacher needs to do to effectively facilitate mathematical discussion, you might list them, in general form, on a flip chart.

TRANSITION | Tell participants that in the next discussion, they will add this additional skill of facilitating mathematical discourse to the learning profiles they have made for teachers in their own school.

FACILITATOR'S NOTES
ACTIVITY 1

About facilitating classroom discourse

In order to make opportunities for children to develop strong conceptual understandings of mathematical ideas, linked to knowledge of facts and procedures, a teacher needs to develop a classroom practice and culture in which children receive support and guidance in talking about their ideas. Children need to come to trust that they will be listened to respectfully and that others will engage actively with their ideas. Not only is this new for many children and teachers, it is also new for many administrators who play a critical role in setting the tone in their schools.

In this activity, administrators examine the nature of facilitating classroom discourse by exploring the strategies of a first-grade teacher, Ms. Scott, as she works with her students in a whole-group discussion. In a classroom such as this one, children need to learn the following:

- how to express their mathematical thinking so that others can understand it
- how to pay careful attention to and understand their peers' mathematical thinking
- how to paraphrase the thinking of other children
- how to build on that thinking

Teachers also need to know and do a lot in order to facilitate discussion that is focused on understanding mathematical ideas. For example, a teachers' skills and competencies should include the following:

- understanding the math
- knowing how to listen for the mathematical meaning of what the children are saying (as in a clinical interview)
- being able to decide which of the children's ideas are the most important to pursue, and then how to pursue them
- being able to ask questions, uncover misconceptions, and pose alternative problems that will help children think about the issue more deeply
- being able to juggle complex classroom management tasks, keep everyone focused, help children understand each other, pull in reluctant children, and attend to the range of learners in the classroom so that everyone is productively engaged
- being aware of the different ways in which the children are engaged in thinking about the problem, even when they are not vocal participants in a discussion

Introducing the video clip

Please review the section Setting a Climate for Viewing Videotapes of Teachers and Students at Work (p. 15) to think through the purposes of videotapes in this course and how you might help participants develop a constructive orientation toward what they see.

This clip is an excerpt from the video series *Relearning to Teach Arithmetic* (further described in the Resource List, p. 134). The filming for this series took place in a range of urban and suburban classrooms, with student populations diverse in their ethnicity, socioeconomic status, and language.

The activity depicted in the videotape is one that many teachers use. However, Ms. Scott takes the activity much further than many teachers might, especially in her choices about what ideas to follow and extend and the way she involves a range of students in the thinking. The discussion takes place in the spring, which means that these children have been learning all year how to work effectively in a discourse-based classroom.

First viewing and discussion

Be prepared for conflicting responses to this video clip. While some administrators might comment on the positive atmosphere in the room and on the wide range of student participation, others might be uncomfortable. They might wonder about the purpose of such an open-ended activity, arguing that it produces a series of problems connected only by their solution. They might find it disconcerting that Ms. Scott seemed just as receptive to children counting on their fingers as to strategies that involved more complex mental arithmetic. Some might be concerned about children's reliance on using fingers and a number line when adding. Others may take the view that children move from these methods to more efficient ones at different rates, and that a teacher should not push children to move from a method that makes sense to them to a more sophisticated one. As always, participants need to listen respectfully to one another's perspectives and to try to understand what underlies the concerns expressed. Be sure that you, as facilitator, solicit and honor all concerns as well as newly forming ideas.

Second viewing and discussion

During the second viewing of the video, the questions on Handout 7 call attention to some of the on-the-spot decisions Ms. Scott makes to help the children in her class articulate their own thinking and listen to that of others.

- **What are some of the strategies Ms. Scott uses to engage a wide range of students in her class?** Ms. Scott asks her students to share a strategy and explain their approaches to adding or subtracting the numbers. She asks them to expand on one another's strategies or to join the class to check on answers given (e.g., skip counting to be sure that she hadn't left out any of the twos that Jason had mentioned).

Her initial question—asking the class to share how they're thinking about numbers and the solutions they came up with, so that they can come up with as many strategies as they can—is open and inviting. Her response to each child's contribution is affirming.

Ms. Scott asks specific questions to draw out children's thinking:

Does anyone remember how we figured out . . . ?

Should we all try to . . . ?

Should we double check . . . ?

What do you notice . . . ?

The way in which Ms. Scott brings other children into the thinking about the 100 – 90 + 6 strategy is illuminating. Perhaps sensing that some children might switch off upon hearing such big numbers, she responded, "100 – 90. Those are big numbers. Who can help me think?"

Norman mentions that he remembers a conversation from a math session earlier in the week in which they found out that 100 – 90 = 10. Ms. Scott asks, "Does anyone remember how we figured that out?"

Nora volunteers that they had counted 10 up from 90. Ms. Scott follows with an invitation for the class to count 10 up from 90, which they do in unison. The sharing continues as Yvonne explains that the next step is 10 + 6, at which point Ms. Scott asks, "Who can help us?"

This episode is a prime example of how Ms. Scott often responds to individual contributions by inviting others to check or extend the thinking.

- **How does Ms. Scott work with the strategies of students who are at different points in their understanding?** A child who can visualize mathematical relationships with large numbers (e.g., 100 – 90 + 6) and a child who is working on patterns in skip counting may be in two different places conceptually. However, both strategies are important in learning mathematical concepts and procedures. In this lesson, Ms. Scott works with all the different strategies. For example, with the strategy that relies on repeated addition or skip counting (2 + 2 + 2 . . .), she draws out Jason's grouping strategy, a step toward understanding multiplication and the concepts of multiples and factors. She also elicits from children the approaches they used to add and subtract the numbers within the strategies they chose. In this short excerpt, children describe counting on their fingers, using the number line, taking apart and combining numbers mentally, and referring to prior mathematical experiences.

- **What are some questions Ms. Scott asks, statements she makes, or other actions that she takes that guide the discussion in mathematically important ways?** Here participants are likely to refer to many of the points already mentioned. In addition, one of the most important things that Ms. Scott does is to write the strategies that children share on a flip chart, so everyone can see visual representations of the verbal descriptions.

Jason's approach of adding eight twos together was a particularly instructive strategy for Ms. Scott to work with *visually*. Ms. Scott asks Serena to check that all the twos from Jason's strategy are included in her recording on the flip chart. Serena counts by twos (skip counting). When Jason explains that he then grouped the twos into fours, Ms. Scott makes a line of four fours. She calls attention to the first line that is all twos, and to the next one that is all fours. Later, when Jason combines the fours into two eights, she points out the repeated pattern, asking, "What do you notice about that line?" By highlighting the number patterns in repeated addition, she is laying the groundwork for future work with multiplication.

From what we can tell from this short segment, Ms. Scott has considerable strengths as a teacher—in terms of her mathematics understanding and the ways in which she works with children's mathematical ideas. There is, however, one error she makes in talking about the first strategy, 100 – 90 + 6. A student comments, "If you count 10 up from 90, you get 100" (this is correct), and Ms. Scott rephrases the student's statement, referring to "how many numbers there are between 90 and 100" (this phrasing is *not* correct: the number of numbers *between* 90 and 100 is 9, not 10). Administrators are unlikely to notice an error like this one during a classroom observation, but if they bring it up, suggest that they think about how they could respectfully point out the issue to Ms. Scott in a post-observation conference.

Notes on keeping the discussion focused

The intent of this activity is to investigate what is involved in facilitating a mathematical discussion. It is important to try to hold the focus of the discussion on the qualities of this kind of discourse and the things both teachers and children need to learn in order to make it work. It can be easy to let the discussion slip into a critique of the teacher. For example, someone might say, "Ms. Scott gave too much wait time," or someone might be critical of the way she defined "good" problems as ones that have "lots of numbers," since there are certainly mathematically interesting examples that would not have lots of numbers. If administrators in your group raise these issues, try turning their critiques into inquiries.

So, what do you think Ms. Scott was trying to do in allowing the room to be silent so long?

What might she have been trying to accomplish when she referred to good problems as ones that have lots of numbers?

This activity can give administrators a practical introduction to the kinds of skills and competencies teachers need to make mathematics discourse come alive in their classrooms and to support the development of students' mathematical ideas. Having some concrete examples can help administrators choose and evaluate the professional development opportunities they provide in their schools and districts.

Relevance for the workplace

You may hear some administrators say that this activity pays too much attention to detail and does not feel relevant to their everyday work. If so, help them consider how this kind of detailed listening and analysis will help them when they observe teachers and do teacher supervision and evaluation. When administrators know what to look for in classes where teachers are changing their practice, they can help to support the teachers' goals by focusing their observations and follow-up conversations on key aspects of standards-based classrooms. Note that it is especially important to make this connection to classroom observation if your group will not be doing *Lenses on Learning, Module 3*.

Some administrators might express frustration if they perceive that their own teachers are far from the image of Ms. Scott. This tape may also raise administrators' concerns about reluctant teachers—teachers who, for different reasons, do not want to take the time or effort to cultivate the skills and competencies needed to facilitate mathematical discussions. While it is important to acknowledge that both limited skill and lack of motivation in teachers present difficult issues, you will want to encourage participants to put these concerns on hold while they consider meaningful professional development designed to address the actual learning needs of teachers. You can mention that the issue of reluctant teachers will be addressed in Session 5.

ACTIVITY 2
Homework Discussion
55 minutes

> **Materials**
> Participant's written homework (teacher profiles)
> Handout 8
> Handout 9 (3 copies per participant)
> Colored pencils or markers

The Teachers in Our Schools

15 minutes	**Adding "facilitating discourse" to the profiles** Ask administrators to turn to the written homework they prepared for this session. Point out that the homework was assigned before they had considered the area of "facilitating classroom discourse." Distribute Handout 8, *Teacher Profiles: Quality of Mathematical Discourse*. Explain that the handout questions are to help them think about the qualities of classroom discourse that support the constructing, experimenting, reasoning, and communication encouraged by the NCTM's *Principles and Standards for School Mathematics*. 1. Does the teacher ask students to explain their ideas and approaches? 2. Does the teacher invite students to respond to each other's ideas? 3. Does the teacher build on students' ideas? 4. Does the teacher engage a range of students in the discussion? Give participants a few minutes to add to their profiles any impressions they have about their three teachers' comfort with facilitating classroom discourse that extends and deepens mathematical understanding.
10 minutes	**Graphing the profiles** Distribute three copies of Handout 9, *Reflecting on the Teachers in Our Schools*, to each participant. They are to create a separate bar graph for each teacher they profiled. Explain that using the same format will standardize the representations, which will be important later in the group discussion of the results. Each bar represents one of the three areas of teacher learning that have been discussed: knowledge of mathematics, capacity to listen to and extend students' thinking, and comfort with facilitating discourse that helps children construct, reason about, and communicate mathematical ideas. Each graph is to show where the administrator thinks that particular teacher is on a continuum of growth.

THE BIG PICTURE
This activity is intended to underscore the idea explored in the first session of this module: For professional development needs, one size does not fit all because the needs of teachers are varied. By talking about the teachers in their schools, administrators reflect further on the link between the ideas in *Lenses on Learning* and the teachers they know. In addition to broadening their perspective on their school community, this activity gives administrators a chance to convey their own reality to their colleagues enrolled in the course and to identify with one another's situations.

10 minutes	**Sharing profiles with partners** When the graphs are finished, ask the group to get into pairs—mixing districts and roles where possible—to share their teacher profiles and graphs. Encourage them to be specific as they describe the strengths and needs of their teachers, giving actual examples when possible. Make clear that the intent is not to label teachers, but rather to begin sorting out their areas of strength and areas in which professional development would be useful.
20 minutes	**Whole-group discussion** Finally, ask participants to think about the implications for their work. You might say, *As you think about these three dimensions of teacher learning in mathematics, consider the impact on your situation. What areas of your work might this knowledge affect? Why and how?*
TRANSITION	Conclude the discussion by saying that participants will next begin to identify characteristics of professional development that can effectively address some of the needs they have seen in their teachers.

Facilitator's Notes
Activity 2

This activity has two purposes. First, it helps administrators link their new understanding of facilitating mathematically rich classroom discourse to teachers in their own schools. Second, it lays the groundwork for considering the various kinds of professional development experiences that different teachers may need.

Adding "facilitating discourse" to the profiles

For homework, administrators were asked to write profiles of three or more teachers in their schools, including some they thought were weak and some they saw as strong in mathematics understanding and making sense of children's mathematical thinking. Before they discuss their findings, they add notes about how well each teacher facilitated classroom discourse during the lesson they observed.

Graphing the profiles

Once they have answered the questions on Handout 8 for each teacher, administrators represent each profiled teacher in a bar graph on separate copies of Handout 9. This exercise provides visual evidence that the competencies that teachers need to teach mathematics successfully are not one-dimensional. Administrators may never have considered that mathematical knowledge, listening to children's thinking, and facilitating mathematical discourse are interrelated in complex ways. For example, knowing the mathematics can contribute to a teacher's ability to discern important mathematical ideas that children are trying to articulate. Furthermore, knowing how to facilitate a mathematical discussion that supports inquiry and extends mathematical understanding may bring out ideas that students are struggling with and are not sure how to express.

Sharing profiles with partners

During this sharing, you may hear comments that it is difficult to categorize a teacher. Emphasize that this exercise is not meant to pigeonhole teachers nor to categorize them in a dichotomous way (strong mathematically/not strong mathematically). Rather, the point is to reflect on the evidence that teachers have different professional development needs. If some administrators feel they cannot make assumptions about how to represent particular teachers because they do not have enough information, acknowledge that this is okay. Encourage them to think of their work as developing conjectures that they will refine and adjust over time.

Administrators may be surprised at the wide variation in the teacher profiles. They may find that some teachers are very strong mathematically but still lack the skills to facilitate a discussion that helps students construct, reason about, and communicate mathematical ideas and concepts. Others may have good facilitation skills, perhaps evident when teaching a different subject, but their mathematical understanding is limited and thus limits the mathematical terrain their discussion can explore. Others may have difficulty listening to students' ideas and therefore entirely miss some opportunities for good mathematical discussion.

As the subtlety and complexity of professional development for teachers begins to emerge, administrators may raise the question of the feasibility of individualized professional development. Be ready to help participants reflect on extremes of "one-size-fits-all" versus individualized professional development. The first situation occurs when districts ask all teachers to take part in the same professional development experience; for example, a district might require all elementary teachers to take the same mathematics course. Individualized professional development, on the other hand, extends a range of options to meet teachers where their needs are, but this approach has financial and time constraints. It also has the downside of limiting "shared experience" and the benefits of that for faculty members. These issues will be revisited in Session 5.

In considering professional development for the skills of facilitating mathematical discourse, some participants may bring up the issue of language-impaired children. This situation requires a different set of skills and calls for teachers to understand the special needs of those particular students. Professional development might require attention to the unique needs of special groups, such as children with learning disabilities or non-English-speaking students.

Whole-group discussion

Being aware of the diversity of their teachers' skills and abilities and recognizing that teachers have varying degrees of sophistication in different areas may help administrators recognize the competencies that qualified mathematics teachers ought to have. When reflecting on areas in which this awareness affects their work, participants might mention the following: how they hire new teachers, how they supervise teachers, how they might help teachers design their own professional development plans, and how they plan the professional development opportunities available to teachers across their school or district.

ACTIVITY 3
Video
55 minutes

Materials
Video clip, *Talking Mathematics: A Teacher Seminar*
Handout 10
Flip chart and markers

Learning Mathematics Together

10 minutes

Introducing the video clip Explain that the video clip, *Talking Mathematics: A Teacher Seminar,* is a nine-minute excerpt of a mathematics exploration that took place in a professional development seminar for elementary teachers. You will show the clip two times, first for a general reaction, and the second time to focus on more specific questions.

Make clear that your intent is not to promote the *Talking Mathematics* program for use in their schools. Rather, your purpose is to convey an image of what professional development for teachers looks and feels like when teachers are encouraged to think and talk mathematically, as members of a community of inquiry.

Briefly explain the mathematics problem the teachers are working on so that participants are not distracted by trying to figure it out.

If you drop a cube into a bucket of paint, six faces would get painted. If you attach a second cube to the first and drop them together into the paint, how many cube faces would get painted?

If you attach a third cube (in a straight line), how many cube faces would get painted? A fourth cube? What's the pattern?

You might write the algebraic expressions of each teacher's solution (from the video) on the board or a flip chart for reference (see About the Video, p. 89).

THE BIG PICTURE
The purpose of this activity is twofold: to extend participants' thinking about the role of discourse in learning and to begin to identify characteristics of professional development that address some of the needs identified in the previous activity.

15 minutes

First viewing and discussion Show the video clip and then ask participants for some general comments and reactions.

What struck you as particularly interesting about this seminar?

How is this seminar similar to or different from your current images of professional development for teachers?

10 minutes | **Second viewing of the video clip** For the second viewing, participants are to focus on the ways that the *Talking Mathematics* seminar has supported the learning of these teachers. Distribute Handout 10, *Talking and Listening in Mathematics*. Suggest that participants use the three questions to guide their second viewing and take notes for reference in the follow-up discussion.

1. What different things about listening to each other's thinking did the two teachers in the video learn?

2. What different things about the role of discourse (or talking about mathematical reasoning while learning mathematics) did these teachers learn?

3. How did they learn those things?

20 minutes | **Whole-group discussion** Bring participants back together for a discussion of the benefits for teachers in the type of professional development exemplified by *Talking Mathematics*. You might ask,

What kinds of learning does Talking Mathematics *make possible?*

How did the structure of this professional development experience enable this learning to happen?

In what ways might a seminar like this one help teachers to bridge the gap between the classroom culture you saw in the Today's Number *video and the traditional classroom culture with which many teachers are more familiar?*

After administrators share some ideas about *Talking Mathematics* as a form of professional development, turn to the questions on Handout 10 to explore these ideas in more detail.

TRANSITION | At the close of the discussion, mention that in the next session of the course, the group will consider additional ways in which professional development can be shaped to address the real needs in schools.

FACILITATOR'S NOTES
ACTIVITY 3

Introducing the video clip

In this activity administrators view and discuss a video that shows teachers in the *Talking Mathematics* seminar exploring and "talking mathematics" themselves.

This video offers an image of professional development which may be quite unfamiliar to some of the administrators in your group. For some, this kind of seminar may be so far removed from their own experience that they have trouble seeing how it connects to their administrative practice. Others may be ready to consider making such professional opportunities available in their districts and will more easily see the implications of this video for their work.

Remind your group once again, in whatever way seems appropriate, about bringing a constructive orientation to the viewing and discussion of the video clip.

About the Video

This video clip offers a glimpse of a form of professional development in which teachers learn by experience that talking and listening play a key role in developing a culture of mathematical inquiry in the classroom. In this excerpt from a teachers' seminar, two teachers discuss how many faces of interlocking cubes are exposed when increasing numbers of cubes are snapped together in a row. The two teachers have explored the patterns that emerged as they added cubes, and they have generalized their observations in algebraic statements that allow them to determine the number of faces exposed for any given number of cubes:

Ms. Lake (Betsy) $\quad 4x + 2 = S$

Ms. Kaeding (Linda) $\quad 6x - (2x - 2) = S$

The video provides a summary of their reasoning. As the video explains, Ms. Lake (in the clip, she is called Betsy) saw the chain of cubes as one whole figure and came up with a generalization that highlighted the whole group of cubes rather than individual cubes. Her generalization can be represented by the algebraic statement $4x + 2$, where x equals the number of cubes. What she saw was that each cube always has four sides exposed and the two end cubes each have an additional side exposed, so that the number of exposed sides equals 4 times the number of cubes plus the 2 exposed ends.

On the other hand, Ms. Kaeding (in the clip, she is called Linda) paid attention to what was happening as each individual cube was added to the others. She noticed that while each cube contributed six sides, the two sides where cubes are joined would be hidden. Her description was summarized by the algebraic statement $6x - (2x - 2)$, where x again equals the number of cubes, 6 equals the number of sides of each cube, and the first 2 equals the number of sides covered. The -2 in the second part of the equation represents Ms. Kaeding's observation that, because the two ends are not covered, she needed to subtract them from the number of sides covered ($2x$).

These two teachers saw the relationship between the number of cubes and sides expressed in different ways. However, while the two explanations seem quite different, the algebraic representations illustrate that they are equivalent: $6x - (2x - 2)$ can also be expressed as $6x - 2x + 2$, which is the same as $4x + 2$.

By "talking mathematics," these two teachers have uncovered the ideas that a mathematical problem can be conceptualized in different ways and there can be different but equivalent mathematical representations of the same situation.

First viewing and discussion

After the first viewing, look for general reactions to the clip.

♦ **What struck you as particularly interesting about this seminar?** Administrators may note the different perspectives the two teachers brought to the same mathematics problem, or they may remark on how involved in the investigation both teachers were. Others may comment on the intellectual community that had formed in the group of teachers or on the depth to which teachers were taking the mathematics.

♦ **How is this seminar similar to or different from your current images of professional development for teachers?** Many administrators will not be familiar with professional development of this nature. What may strike them as novel is the way the seminar was structured to enable meaningful teacher learning to take place. Notable characteristics of this professional development experience include the following:

- the fact that participants worked with engaging mathematics problems and activities
- the opportunity participants had to do and "talk" mathematics with each other
- the way in which the facilitator in the seminar listened intently to participants' discussions
- the fact that participants were given ample time to think and talk about the problems
- the fact that involvement in the *Talking Mathematics* seminar is a long-term commitment, in which teachers deepen their understanding of the mathematics and pedagogy over time
- the encouragement participants received to share the range of ways in which they thought about the mathematics in the problems
- the fact that the mathematics participants did was connected to classroom work

Second viewing of the video clip

Following the second viewing, focus discussion on the questions on Handout 10.

♦ **What different things about listening to each other's thinking did the two teachers learn?** Listening to each other's thinking and articulating their own thinking helped the teachers in this seminar gain an awareness of the important role that reflective or investigative talk plays in learning mathematics. Another point to bring up, if administrators don't, is the important role the facilitator plays in drawing out central mathematical ideas from the teachers' work—parallel to the role a teacher plays in drawing out key mathematical ideas from children's work.

♦ **What different things about the role of discourse (talking about mathematical reasoning) did these teachers learn? How did they learn those things?** The experiences teachers had in this seminar—doing the mathematics for themselves and understanding the ways other people are thinking about the mathematics—provides an image of the kind of culture it is possible to create in a mathematics classroom. The experience underscores the power of a structure in which students are working to understand the mathematics for themselves as well as working to understand the thinking of others and, in the process, constructing a deep and rich understanding of the mathematics.

Whole-group discussion

The purpose of this discussion is to help administrators generalize about the nature of the professional development depicted in the video.

♦ **What kinds of learning does *Talking Mathematics* make possible?** Drawing on their notes to the questions on Handout 10, participants may comment on the attentive listening the teachers did and on the role of discourse in mathematical learning. They may also note the following: In a seminar like this, teachers learn some important mathematics; they experience for themselves what it feels like to construct their own understanding in a mathematics class; they observe what goes into facilitating a mathematics investigation such as this one; and they learn how to listen to and learn from each other's approaches.

◆ **How did the structure of this professional development enable this learning to happen?** Participants should note that teachers worked individually as well as in collaboration with others; they articulated their strategies to each other and to the facilitator, who helped them to situate their thinking in a larger framework; they shared their thinking with other members of the group; and they worked to understand each other's approaches, thereby enriching their own.

◆ **In what ways might a seminar like this one help teachers to bridge the gap between the classroom culture you saw in the *Today's Number* video and the traditional classroom culture with which many teachers are more familiar?** One of the teachers was surprised that her partner for this activity saw how to solve the problem right away while she herself needed more time. This same teacher's way of thinking about the problem differed from that of her partner, and she realized that her own thinking was greatly enriched when she made the effort to listen to and comprehend her partner's thinking. Being in a seminar such as this one makes it possible for teachers to experience for themselves what goes into a classroom culture like the one shown in the *Today's Number* video.

Your role as facilitator

During this activity, there are many ways to help administrators make connections to their own work. This may be your most important role as you listen to the participants in the whole-group discussion.

Help administrators see the connection between this seminar for teachers and what teachers can do in their classrooms. Specifically:

- Teachers who discover the different ways their colleagues think about a problem may become more open to the range of ways their students approach problems.

- Teachers who realize that their own thinking was enriched by listening to the thinking of others might be motivated to help students focus more on listening to each other, rather than on getting to the answer quickly.

A seminar like this one can be very instructive for teachers in the way it models a classroom experience. While teachers may not immediately be able to replicate their own experiences in their classrooms, they will have an image of where they would like to be heading.

You may hear the administrators in your group express a concern that the mathematical talk is too sophisticated for their teachers. If this is the case, remind participants that changing one's classroom practice is a multifaceted process that takes place gradually. The teachers in this video may be further along in the process than the teachers in their schools currently are, but when these same two teachers first started the process, they may have been much less able to articulate their ideas and understand each other's thinking.

By monitoring such perceptions, you can help administrators make connections between the image of professional development in this video and the image of classroom discourse in the earlier clip.

Administrators sometimes make connections between the reflective community depicted in the *Talking Mathematics* video and their own work together in the *Lenses on Learning* course. If this happens, take this opportunity to revisit the principles of reflective communities that underlie your work together. If no one raises this point, you might ask if they noticed parallels between what the teachers in the video clip were doing and what they are doing in this course.

Tell participants that in the next two sessions they will be exploring further images of professional development. Suggest that they reflect on the ideas that they've explored so far and how they relate to the professional development that they provide in their own districts. While it would be premature to launch a full discussion of their individual professional development approaches, it is not too soon for administrators to begin thinking about this on their own.

CLOSING
15 minutes

Bridging to Practice

At the close of this session, invite participants to respond in their journals to the Bridging to Practice questions:

1. Pick an idea that came up today that you found particularly interesting. What is your current thinking about this idea?

2. Where is your school now with regard to this idea?

3. What are one or two things that you will go back and pursue, to move yourself and/or your school along with this idea?

Homework

Distribute Handout 11, *Homework for Session 3*. The assignment is Reading 4, Brian Lord's article "Teachers' Professional Development: Critical Colleagueship and the Role of Professional Communities." This reading contrasts traditional approaches to professional development to a model, termed "critical colleagueship," that cultivates greater reflectiveness and sustained learning among teaching colleagues.

Discussing the *Today's Number* Video

These expressions summarize three mathematical strategies used by children in Ms. Scott's first-grade class to "make" the number 16.

100 − 90 + 6 100 − 90 = 10 10 + 6 = 16	2 + 2 + 2 + 2 + 2 + 2 + 2 + 2 = 16 Skip count by 2s. Combine series of 2s to get four 4s, and then two 8s, and then one 16.	8 + 1 + 2 + 1 + 4 8 + 1 = 9 2 + 1 = 3 9 + 3 = 12 12 + 4 = 16

DISCUSSION QUESTIONS

Keep these questions in mind during your second viewing of *Today's Number*. Take notes for reference during the follow-up discussion.

1. What are some of the strategies Ms. Scott uses to engage a wide range of students in her class?

2. How does Ms. Scott work with the strategies of students who are at different points in the development of their understanding of mathematical patterns and relationships?

3. What are some questions Ms. Scott asks, statements she makes, or other actions that she takes that guide the discussion in mathematically important ways?

Teacher Profiles: Quality of Mathematical Discourse

For homework, you wrote profiles of several teachers in your school. You focused on their skills in understanding the mathematics they teach and making sense of their students' mathematical thinking. Now, consider each teacher's ability to facilitate good mathematical discussions in class. Add a section to each teacher's profile that answers the following questions, to the best of your knowledge.

1. Does the teacher ask students to explain their ideas and approaches?

2. Does the teacher invite students to respond to each other's ideas?

3. Does the teacher build on students' ideas?

4. Does the teacher engage a range of students in the discussion?

Reflecting on the Teachers in Our Schools

Teacher: _____

STRONG

WEAK

| Teacher's mathematical knowledge | Capacity to listen to children's thinking | Comfort with facilitating discourse |

KEY

RED	Teacher's mathematical knowledge
BLUE	Capacity to listen to children's thinking
YELLOW	Comfort with facilitating discourse

Talking and Listening in Mathematics

As you view the video clip *Talking Mathematics: A Teacher Seminar* for a second time, take notes for a discussion of these questions.

1. What different things about listening to each other's thinking did the two teachers in the video learn?

2. What different things about the role of discourse (or talking about mathematical reasoning while learning mathematics) did these teachers learn?

3. How did they learn those things?

Homework for Session 3

Before the next session, please consider the following:

READING 4 "Teacher's Professional Development: Critical Colleagueship and the Role of Professional Communities" by Brian Lord

We will begin the next class session with an in-depth discussion of this reading. Using this as a starting point, we will then look into alternative forms of professional development.

As you read, take notes about ideas in this article that strike you. In particular, think about the distinguishing characteristics of the two modes of professional development the author describes.

SESSION 4

Critical Colleagueship

The profession of teaching is changing. Teachers were once considered isolated experts in their classrooms; now they are called on to build sustained communities of colleagues, working together to understand the hard and complex issues of teaching and learning. Colleagueship, in this sense, has become an important conduit for teachers' professional growth. In order for colleagueship to function in this way, teachers need to learn to be both empathetic and intellectually critical. A school culture in which "critical colleagueship" thrives, as described in Reading 4, is able to "support teachers in their efforts to bring to the surface [their] questions and concerns, to help teachers expose their classroom practices to other teachers and educators, and to enable teachers to learn from constructive criticism" (*Module 2* readings, p. 70). Such a school culture finds strength in the diversity of backgrounds and experiences among the faculty.

The kinds of professional development for teachers that support a collaborative school culture have a different theoretical framework from more typical forms, which tend to be short-term experiences addressing issues and questions that are externally defined. Newer approaches, reflecting NCTM's *Principles and Standards for School Mathematics,* recognize the complexity of the challenges inherent in changing classroom practice and the longer amount of time needed for those changes to take place. Newer approaches also recognize the value of teachers working together, both to define the issues they want to address and to search for answers.

This session lays out the theoretical foundations for new approaches to professional development, introducing a variety of possible approaches. These kinds of programs, when tailored to the needs of particular teachers and their settings, can be important components of professional development in a school community.

OVERVIEW FOR SESSION 4

OPENING page 102 5 minutes	**INTRODUCING THE SESSION** Begin with announcements and a brief preview of the session agenda.
ACTIVITY 1 Homework Discussion page 103 85 minutes	**EXPLORING CRITICAL COLLEAGUESHIP** Participants review Brian Lord's article (Reading 4) and consider ways in which a culture of critical colleagueship in their schools can set a climate for learning among adults as well as children. They explore the significance of this new image of the profession of teaching as described by Lord and discuss how such a culture might be established in their own schools and districts.
ACTIVITY 2 Readings page 107 75 minutes	**ALTERNATIVE IMAGES OF PROFESSIONAL DEVELOPMENT** This activity introduces five different approaches to professional development for teachers that address the kinds of needs they have been exploring in this module and that build on the climate of critical colleagueship. Working first in small groups and then as a whole group, administrators collectively reflect on these approaches, both from the perspective of their own understanding and in light of Lord's view of critical colleagueship.
CLOSING page 112 15 minutes	**BRIDGING TO PRACTICE** Participants finish the session with guided reflective writing that helps them link the ideas in Session 4 to their own work as administrators. **HOMEWORK** For the next session, participants complete two readings and write up a memo about professional development in their own communities.

BIG IDEAS	This session explores • the differences between the training paradigm and "critical colleagueship" as alternative approaches to professional development • different kinds of professional development that build on critical colleagueship • the promises and tensions inherent in professional development that embodies critical colleagueship
MATERIALS	☐ Flip chart and markers ☐ Readings 4–9 for *Lenses on Learning, Module 2* ☐ Handout 12, pages 113–114 ☐ Bridging to Practice display chart or transparency
PREPARATION	• Prepare and post an agenda for the session, noting the time allotted for each activity. • Read the article assigned for homework in the previous session, Reading 4, "Teachers' Professional Development: Critical Colleagueship and the Role of Professional Communities." In Activity 1, administrators will draw up a list of the central characteristics of the traditional training paradigm and critical colleagueship. Before the session, make such a list yourself to become familiar with the ideas and to anticipate points that administrators may bring up. Think through each of the subsequent discussion questions (listed in the activity) for the same purpose. • For Activity 2, become familiar with the five approaches to professional development outlined in Readings 5–9. Think through the characteristics they have in common and the ways in which they are distinct from one another.

OPENING
5 minutes

> **Materials**
> Posted agenda

Introducing the Session

5 minutes

Session preview After announcements, give a preview of today's agenda, explaining that the session breaks into two parts. This first half will be a thorough discussion of the article by Brian Lord (Reading 4), which they read as homework. During the second half, participants will work primarily in small groups to explore some alternative possibilities for professional development as put forth in Readings 5–9.

Taken as a whole, today's session gives administrators a chance to step back and examine theoretical frameworks for professional development and to consider a range of strategies that build on critical colleagueship.

As this session progresses, you will want to be particularly attuned to participants' evolving perspectives on teacher learning. Notice which words and images come to the surface as individuals work with the theoretical analysis of professional development.

ACTIVITY 1
Homework Discussion
85 minutes

> **Materials**
> Reading 4
> Flip chart and markers

Exploring Critical Colleagueship

20 minutes	**Reviewing the article** Post two sheets of flip-chart paper, one headed *Dominant Paradigm,* the other headed *Critical Colleagueship.* Ask participants to suggest characteristics of each mode of professional development. Record each idea on the appropriate list. Ask others to share their thinking about each characteristic as it comes up. Avoid displaying a preference for either model.
25 minutes	**Comparing and contrasting forms of professional development** Discuss the possible effects of each mode on teachers, considering how and what they might learn. Start by inviting participants to talk to a partner; after a few minutes, open the discussion to the whole group. *What are teachers likely to learn from each of these two modes of professional development?* *What impact might that have on students' learning?*
20 minutes	**Supporting collegial colleagueship** Next, ask the group to think in terms of taking "critical colleagueship" into their own practice. Again start in pairs, then move to a whole-group discussion. *What policies, practices, and structures would support critical colleagueship in your schools and districts?* *How might links to the kinds of "resource-rich professional communities" mentioned in Reading 4 support these efforts?* **THE BIG PICTURE** Resource-rich communities can broaden the capacities of already vital schools, bringing in perspectives that are not found within their own faculty. This discussion helps administrators consider how connections with a wider community can enrich what happens within their schools.
20 minutes	**Discussing challenges** Finally, discuss the difficulties they might face. *What challenges might there be in trying to institute the policies, practices, and structures that support critical colleagueship?*
TRANSITION	Tell participants that they will next explore some alternative possibilities for professional development that may be new to them.

Facilitator's Notes
Activity 1

Reviewing the article

The purpose of this activity is to help administrators begin to tease out differences between traditional approaches to professional development, which are based on a "training" paradigm, and approaches in which teachers discuss substantive issues about teaching and learning with colleagues, both in their own school and beyond.

In Reading 4, Brian Lord argues that what he calls the "dominant paradigm" for professional development (a training paradigm) cannot provide teachers with what they need to transform their teaching for the following reasons: (1) It communicates a restricted view of teaching; (2) it "tells" teachers what to do, rather than providing opportunities for exploration; and (3) it delivers too small a toolbox, limited to behaviors, skills, and items of knowledge, rather than conceptual sophistication and sound professional judgment under conditions of uncertainty.

Lord contrasts this with a new collegial view, in which teachers participate in sustained self-reflection and dialogue, think critically, are open to new ideas, and are empathetic toward colleagues' dilemmas.

The first discussion in this session reviews the characteristics of the dominant paradigm and of "critical colleagueship." A set of typical lists that administrators have generated is shown below. Your group may come up with somewhat different lists, but the main contrasts between the two modes should emerge. Facilitators should avoid revealing a preference for either model. You need to welcome voices that support the traditional approaches as well as those that embrace critical colleagueship.

While *Lenses on Learning* itself is based on the premise that more professional growth can be accomplished through a collaborative approach, there are times when traditional approaches accomplish what is needed. Administrators need to develop a sense of what is appropriate in different circumstances.

Dominant Paradigm	Critical Colleagueship
Short-term—it doesn't take long to understand the problem	Long-term—problems take a long time to be understood and worked on
Transmission model—instructor gives information to participants	Teachers search for understanding together and appreciate their differences
Individual teachers, one by one	Collaborative work among teachers
There's an external definition of the questions and issues to be dealt with	Teachers define their own issues from their own experience
Delivers the answers to teachers	Allows for uncertainty and questioning; the teacher comes to some sort of conclusion
Does not model the way we want teachers to work with students	Models the way we want teachers to work with students

Comparing and contrasting forms of professional development

For the most part, the prevailing view of professional development for teachers does not give them opportunities to construct their own understanding. Instead, they are introduced to an idea with the expectation that they will then move it into their practice—all in short order. This approach can work well to communicate straightforward information, such as ways to physically set up a classroom to support more discussion among students. However, the training model is less satisfactory for addressing the complex matter of changing one's classroom practice. Through critical colleagueship, on the other hand, teachers can learn how their colleagues view issues that are important to the group as a whole, while they articulate their own evolving perspective. This mode assumes that learning is an ongoing process and that teachers may need to work on any given issue over a long period of time.

The question of the impact of a critical colleagueship approach on students' learning echoes a theme from Session 3. In the *Talking Mathematics* video clip, teachers' participation in that seminar provided a living image of the culture it is possible to create in a mathematics classroom. Professional development delivered through the training paradigm, on the other hand, does not model the approach recommended for teachers who are exploring and strengthening students' mathematical ideas.

Critical colleagueship in your group

By now, group members are probably comfortable talking together, sharing their ideas about mathematics learning and teaching and the implications of new ideas for their own administrative practice. Norms for exploring ideas openly and thoroughly, listening respectfully, and trying to understand one another's ideas are by now well established. Trust among participants has been growing as they experience the nonjudgmental and accepting atmosphere that you have created in the classroom.

Critical colleagueship suggests just such community. While discussing Reading 4 in terms of the culture in their schools, some administrators may notice that they have been building this type of community in the *Lenses on Learning* class. That is an intentional aspect of the design of this course.

If participants mention a relationship between the critical colleagueship that they might foster for teachers and the way their seminar group has been working for them, discuss with them what has made it possible for the group to work in this way. Using their own experience in the seminar as a reference point, administrators can better understand how critical colleagueship can function for teachers. They should recognize that their different backgrounds and values are assets to the group's evolution. They might also explicitly discuss the norms and practices that have made the group a place for earnest searching and supportive conversation, understanding that these same norms and practices will be needed to create critical colleagueship for teachers.

Administrators who have noticed the parallels have observed the following:

I'm trying to think about what it means to be "collegial." For example, this [seminar] group represents an effort at collegiality. I feel that we are intellectual colleagues here. I feel that we're tacitly agreeing to deal with difficult questions in a very open way.

———

I think there's something to be said about the development of "groupness." It would be interesting to have someone join this group at this point, because over [the time we've been meeting together], we really have grown into this relationship with each other. We certainly didn't start there. Critical colleagueship must be something that develops and grows. And it is fostered by talking together about things that really matter to us.

Who's in the room matters. They need to be people who respect each other, who are trusted, who are seen as having good ideas and willing to share. And each group would be different. I mean, it's the composite of this particular group of people that makes this particular set of critical colleagues. It would be different with a different bunch of people.

Supporting collegial colleagueship

This session is rich in potential connections to administrators' own practice. If they make the connection between the way the group itself works and critical colleagueship for teachers, they should see that they can use their own experiences of learning—in this group and elsewhere in their practice—as sources of ideas for thinking about learning and teaching more generally. Many express a desire to take critical colleagueship directly into their practice. As one participant said,

In what ways can I explore, or share, or encourage, or demonstrate the collegial model in my practice? This is a real question I will be mindful of as I work with my department.

A number of practical issues may arise, as evidenced in the following musings:

I wonder what the effect would be to have teachers in my school read this article and meet for an extended period of time to discuss it. (An hour after school won't do.)

How can we structure time in the teachers' contracts, in the budgets, in the school calendar, to provide the opportunity to engage in critical colleagueship?

In this discussion, administrators should also consider the role of "resource-rich professional communities"—groups of colleagues beyond the school and district that can broaden their capacities, enriching what happens within the school and bringing new perspectives.

Discussing challenges

In the closing discussion, administrators may mention the amount of time it takes to build the trust that critical colleagueship requires. They may speculate about their own behavior on the job—do they function in a critical colleagueship mode?—and about what it means to support teachers' efforts to develop colleagueship, rather than take charge of the process themselves.

Some administrators may speak of a general unwillingness among staff developers and teachers to let go of the prevailing training paradigm. This paradigm is familiar and provides many things they need, whereas it is not necessarily clear how to make critical colleagueship work. Some administrators in the group may observe that the training paradigm is good for some kinds of teacher learning and critical colleagueship is good for other kinds. Others may bring up the discomfort that comes from having ideas challenged in the critical colleagueship mode and wonder if this is appropriate for all teachers. It is important to explore all such thoughts as participants sort through what it might be like to move their schools and districts in the direction of critical colleagueship. A key point in this discussion is to recognize that a district's current policies, procedures, and indeed the very structure of the school day may inadvertently support only the dominant paradigm and make it difficult to explore critical colleagueship.

ACTIVITY 2
Readings
75 minutes

Materials
Readings 5–9
Highlighters and stick-on notes
Flip-chart paper and markers for each small group

Alternative Images of Professional Development

20 minutes	**Individual reading** Explain that while the type of collegial culture described in Lord's article is extremely important in setting a climate for learning among adults in schools, equally important are the types and quality of the professional development opportunities that we make available to teachers. This next activity introduces several complementary approaches to professional development. Administrators are asked to reflect on these offerings, both in light of their own experience and circumstances and in relationship to Lord's view of critical colleagueship and the use of resource-rich professional communities.
	Form groups of five participants. Each group member is assigned one of the five readings about a particular professional development offering.
	Distribute highlighters and stick-on notes for use in marking key points. Remind participants that they should expect to read at different rates; urge them to take the time to digest what they read. Each individual should be prepared to present his or her assigned professional development strategy, sharing what seems beneficial and what seems potentially problematic about the approach.
30 minutes	**Article review and small-group discussion** Give each group flip-chart paper and markers to record the main points of their discussion. Group members should first take turns sharing the features of all five approaches, recording key ideas on chart paper. Then, as a group, they should consider the questions on page 87 of the readings.
	What characteristics do these strategies for professional development have in common?
	How are they distinct from one another?
	What promises and what tensions do you see in these approaches to professional development?

Alternative Images ◆ 107

THE BIG PICTURE
The list of tensions that participants develop in each small group acknowledges the constraints within which administrators are working. The benefits represent the very real openings that are possible when teachers engage in substantive professional development experiences.

25 minutes

Whole-group discussion Ask each group of five to share with the whole group; together they should cover the main points of their small-group discussion. Try to draw out the idea that a school culture of critical colleagueship enhances both teachers' capacity to engage in such professional development activities and the school's capacity to provide them.

As a link to administrators' own practice, ask them to brainstorm a collective list of questions they will want to keep in mind when they are considering professional development programs for their schools. You could ground the brainstorming in a specific situation, as follows:

Imagine that you are in the process of choosing a summer professional development program to recommend to the teachers in your school. You have in front of you brochures from three different programs, all of which say that they are designed to help teachers work with standards-based mathematics.

When you call for more information about these programs, what questions might you want to ask?

What information or guidance might you seek from teachers or teacher leaders in your district to help you assess these professional development options?

TRANSITION

Explain that in the final session of this module, participants will expand their thinking about professional development, not just as something that meets individual needs but also meets the goals of the wider school or district community.

Facilitator's Notes
Activity 2

Readings 5–9 describe actual programs that existed at the time of this writing, giving administrators a sense of what can be done, even if the programs are not currently available in their own communities. One goal here is to enlarge administrators' vision about how professional development can work.

Article review and small-group discussion

During the small-group discussions, circulate and listen to the ways participants are reflecting on the five possibilities for professional development. What potential for teacher growth do they see embedded in each structure? What types of obstacles do they bring up? Do they refer to the ways in which a collegial school culture might support teachers who are making substantial investments of time and reflection? Also, listen to the ways each group addresses the questions about the five approaches as a whole.

- **What characteristics do these strategies for professional development have in common?**

Characteristics common to most of the strategies include these: (1) teachers deepen their mathematics knowledge; (2) teachers learn in situations that parallel the pedagogy they are striving to bring to their classrooms; (3) there is an emphasis on understanding the diversity of children's thinking; (4) each professional development experience is ongoing, extended over time; and (5) each approach assumes that collegial relationships are central to the learning experience.

- **How are they distinct from one another?**

Participants are likely to point out that a "professional network" is different from the other four strategies in terms of the nature of teacher engagement. Rather than exploring with immediate colleagues ideas that are grounded in their day-to-day classroom practice, teachers join professional networks in order to broaden their perspectives beyond the immediate experience of their classroom or school community. This strategy reflects Lord's articulation of the importance of being connected to resource-rich professional communities, in addition to being grounded in a community of critical colleagueship.

Professional networks are further distinguished from other forms of professional development by the capability of members to electronically access information, participate in course work, and stay connected. This capability provides enormous flexibility and allows for initiative to be taken by individual teachers as well as by groups of teachers within a school or district.

The forms of professional development also differ in the aspect of a teacher's practice each one targets. For example, *Immersion in Inquiry into Science or Mathematics* helps teachers to deepen their mathematics knowledge, whereas *Examining Student Work and Student Thinking and Scoring Assessments* helps teachers increase their understanding of students' thinking by looking at student work.

During this discussion, administrators may bring up forms of professional development other than those mentioned here. For example, one powerful vehicle for professional development is *supervision,* particularly when provided in conjunction with some of the approaches described in Readings 5–9. While the other approaches offer focused or intensive work on specific areas (e.g., mathematical exploration or working with new curricula), supervision provides the situated, one-on-one interactions that can help teachers bring these new ideas alive and give them concrete meaning within their own particular classrooms. During pre- and post-observation conferences, when administrators and teachers reflect together on the events that took place during a class session, administrators can support teachers as they construct their own understanding about mathematics learning and teaching. In this process, administrators as well can be building new knowledge that will inform decisions they need to make in their practice.

Another idea mentioned but not specifically featured in the readings is the development and support of *teacher leaders*. This is another powerful way of helping teachers to grow professionally. The administrator's first task is to identify staff members who are in a position, by virtue of their understanding of mathematics and their knowledge of children's thinking and productive ways of building on their ideas, to take on the responsibility of working with other teachers. Beyond identifying these individuals, administrators need to help these leaders become effective facilitators of growth among other adults in the school, and they need to develop contexts in which this leadership cadre can extend their own learning over time.

◆ **What promises and what tensions do you see in these approaches to professional development?** These efforts promise to deepen teachers' mathematics knowledge and their knowledge about their students' mathematical thinking, and they engender a shared vision of mathematics teaching and learning.

One tension that participants will likely mention is limited resources (teacher time, teacher leadership, access to outside resources, financial resources, and so on). Because such constraints are real and constant, it is important to make room for them to surface, even as new possibilities are being explored. Try not to let this turn into a gripe session; instead, it should be an opportunity to map out the issues administrators need to tackle. When participants raise a type of constraint, ask them to be specific and describe it in more detail. In the process of thinking through the specifics of limited teacher time, for example, they might generate some possible ways to change things.

Administrators may also bring up their historical relationship with teachers as potentially interfering with collaborative work. Specifically, the issues of evaluation and tenure can dissuade teachers from openly sharing their concerns, weaknesses, and hopes for their own practice. Yet, sharing these with colleagues is an important part of constructing new understandings of teaching practice. Administrators will have to consider how they can ensure that they do not penalize teachers for sharing their concerns as part of their professional growth. If this issue comes up, you can liken teachers' sharing to administrators' own sharing about their uncertainties in this reflective community. Thinking about how they would like others in the group to respond to *their* sharing may suggest how they in turn might respond when teachers confide in them.

Lead the group to see ways in which the professional development possibilities presented in Readings 5–9 might help to address certain constraints. For example, if they report a lack of teacher leaders in their district, they might find that professional networks could enhance the quality and number of teacher leaders.

Make note of the specific challenges administrators anticipate in their own situations and add these to the list of concerns you started in Session 1. You will use this list in Session 5.

Whole-group discussion

As administrators think about the additional information they might want to obtain about professional development options, they should consider questions like these:

- "How often does it meet and for how long?" Knowing this would give a clear sense of how the program developers think about professional development—as a one-shot deal or as a process that takes time to develop.

- "What aspects of a teacher's practice does each professional development experience target?" This question reflects an understanding that different programs will meet the needs of different teachers.

- "What are the activities in which teachers engage, and how are they run?" Administrators need to know what model of professional development the program follows: Is it a transmission-of-information model, or one in which teachers define their own issues and search for understanding together?

Since teachers have differing professional development needs, seeking their input would help an administrator decide which programs make the most sense and for which teachers. Administrators might share Readings 5–9 to give their teachers and teacher leaders some concrete images of the kinds of experiences that are possible and available.

This discussion will give you a good idea of how administrators are thinking about the issues they have been addressing during this *Lenses on Learning* session. What views of professional development do their comments reflect? Do they see that teachers themselves need to be part of the process of selecting appropriate professional development experiences?

Your own journal writing

While participants are doing their Bridging to Practice reflective writing at the end of this session, you have much to reflect on yourself. Since you are coming to the end of the module, it would be useful to review your previous journal entries and the earlier journal writing of the administrators in your group. Take stock of where you are now in the journey of thinking about professional development for teachers. Of the many ideas about professional development for teachers that have been discussed, which still seem to be open and problematic for the administrators in your group? To what degree has critical colleagueship developed in this group?

CLOSING
15 minutes

Bridging to Practice

Invite participants to respond to the Bridging to Practice questions:

1. Pick an idea that came up today that you found particularly interesting. What is your current thinking about this idea?

2. Where is your school now with regard to this idea?

3. What are one or two things that you will go back and pursue, to move yourself and/or your school along with this idea?

Today's reflective writing can get administrators started thinking about professional development in their own community, a subject they will further address in their homework.

Homework

Distribute Handout 12, *Homework for Session 4,* and be sure participants understand that there are three parts to the homework they will do in preparation for the next session: two involve readings, and one entails reflecting and writing about professional development in their own communities.

For Readings 11–13, assign the readings in any way you wish, being sure that several people will have read each one as background for discussing possible refinements of Project IMPACT in the next session. Everyone should be prepared to discuss one of the three selections.

Homework for Session 4

Project IMPACT

Read the following article about Project IMPACT, a professional development project in mathematics that was conducted in Maryland.

READING 10 "Empowering Children and Teachers in the Elementary Mathematics Classrooms of Urban Schools" by Patricia Campbell

As you read, think about the benefits and trade-offs there might be in attempting to reach *all* teachers in a district with a common set of professional development activities.

Professional Development at Your School

Write a memo about the professional development activities that teachers in your school have been involved in. Include all professional development activities for mathematics; you may include other subjects if you like. What learning was at the core of what was addressed in these activities? Did any of them help teachers develop their capacity to understand the subject itself (mathematics or other), listen to children's understanding of the subject, and facilitate discourse?

This is an opportunity to take stock of the recent history and patterns of professional development in your school. Base your memo on data from your school—records, memos, correspondence, conversations. This memo must be handed in, but it will not be shared with other course participants without your permission.

Readings 11–13

Read one of the three articles on professional development described on the next sheet. Be prepared to discuss the questions in class. (You will be assigned one reading that you will be responsible for discussing.)

Read one of the following, as assigned.

READING 11 "Teacher Support" by Lynn Goldsmith, June Mark, and Ilene Kantrov

Reading 11 is taken from a guide designed to support school districts as they select and use new instructional materials aligned with the NCTM *Standards*. Although the introductory paragraph assumes an audience that has already chosen a curriculum, the reading is equally helpful for schools at different stages of the process. Reflect on the following questions:

- How have practical and conceptual elements of professional development been balanced in this reading?

- How does the vision for professional development laid out in this reading help teachers move from logistical questions of implementation to ideas that are focused on student learning?

READING 12 "Learning Mathematics While Teaching" by Susan Jo Russell, Deborah Schifter, Virginia Bastable, Lisa Yaffee, Jill Bodner Lester, and Sophia Cohen

This essay is from a collection that explores issues in the transformation of mathematics teaching. Reflect on the following:

- How were the teachers described in this paper able to develop their own mathematical understanding in the classroom?

- How would you characterize what has made this learning possible?

READING 13 "Improving the Quality of On-the-Clock Professional Time: What Can the Innovative U.S. Schools Teach Us?" by Nancy Adelman

Reading 13 is from a study that compares the quantity, structure, and use of teachers' time in elementary schools in the United States, Germany, and Japan. In particular, it examines the time when teachers are not in direct contact with students—the time spent planning, preparing, assessing, and collaborating with colleagues. This chapter looks at a number of innovative schools and their efforts to restructure the use of time to create more effective learning environments. Reflect on the following:

- How have flexible approaches to time strengthened the learning culture for adults in the innovative schools described in this reading?

SESSION 5

Providing Professional Development

Throughout Sessions 1–4 of this module, administrators have been broadening, deepening, and enriching their understanding of professional development for teachers. They have reflected on how to support individual teachers' growth in mathematics teaching and learning, and they have begun to construct new images for professional development for themselves. In this final session, they consider what it means to bring these ideas into entire schools and districts that are moving toward standards-based mathematics.

The ideas of the *Lenses on Learning* course take on new meaning when seen from the perspective of the schools or districts for which the administrators are responsible. This session gives participants a chance to stand back and look at the promises and the tensions that emerge when we try to bring together our ideas about individual teacher growth and larger-scale efforts to move entire systems forward.

In time, administrators will resolve these tensions for themselves and find ways to balance the need to be responsive both to individual teachers and to school- or district-wide priorities. For now, this is a valuable opportunity for them to recognize, articulate, and reflect on these tensions with peers, in a seminar setting where time and space have been set aside to do so, while the ever-present demands for decisions or plans are, for the moment, put on hold.

Overview for Session 5

OPENING page 118 5 minutes	**BRINGING IT ALL TOGETHER** As *Module 2* draws to a close, administrators review the ideas covered thus far and then get a preview of this final session.
ACTIVITY 1 Homework Discussion page 120 35 minutes	**PROFESSIONAL DEVELOPMENT IN OUR SCHOOLS** Equipped with new ideas about professional development, participants share the reflections they prepared for homework in which they took stock of the professional development activities in their own schools and districts.
ACTIVITY 2 Readings page 122 60 minutes	**VIGNETTES OF INDIVIDUALS MOVING IN THE STREAM** Administrators turn their attention to four specific examples of teacher learning. They see the range of experiences and influences that bear upon individuals' learning trajectories, and they see how that learning can include some of the professional development experiences that were explored earlier in this module.
ACTIVITY 3 Readings page 125 65 minutes	**A SYSTEM ON THE MOVE: PROJECT IMPACT** Administrators discuss and think further about Project IMPACT, the model of professional development for supporting standards-based mathematics that they read about for homework in Reading 10. They consider how this plan might be adapted to meet the needs of a wider range of teachers.
CLOSING page 129 15 minutes	**BRIDGING TO PRACTICE** Participants end the session with guided reflective writing that helps them link the ideas in Session 5 to their own work as administrators.

BIG IDEAS	This session explores • the relationship between the learning a teacher needs or wants to do and the range of professional experiences that might provide that learning • the promises and tensions that emerge when trying to meet the needs of individual growth and, at the same time, move entire systems forward
MATERIALS	☐ Flip chart and markers ☐ Readings 10–14 for *Lenses on Learning, Module 2* ☐ Participants' written homework from Session 4 ☐ Bridging to Practice chart or transparency
PREPARATION	• Prepare an agenda for the session that lists the major activities and times allotted for each, and post it where it will be visible to all. • On flip-chart paper or an overhead transparency, prepare a list of the challenges participants have identified in previous sessions as they thought about applying the ideas of this course to their own school systems. (You may have made note of particular problems mentioned in Session 1 and Session 4.) • In preparation for Activity 2, become familiar with the vignettes in Reading 14. In particular, think through for yourself the influences on these teachers' growth and the ways in which each individual in turn influenced the growth of others. • In preparation for Activity 3, review Reading 10, which describes Project IMPACT. Think through what is gained and lost in a large-scale effort such as this one, in which a single program attempts to reach all teachers in a district. How might you imagine adapting such a program to incorporate some of the understandings about teacher learning that have developed since 1989?

OPENING
5 minutes

Materials
Posted agenda

Bringing It All Together

5 minutes

Module review After a general welcome and announcements, mention that this is the last session of the module on professional development for teachers. Briefly run through the major topics that have been discussed in this module.

- Changing one's teaching is not just a matter of learning new techniques, but involves changes in knowledge and beliefs as well.

- Teachers need to learn new orientations and new skills—especially developing their mathematical understanding, learning how to listen to children's mathematical thinking, and learning to facilitate mathematical discourse in the classroom.

- There are many new and innovative ways to provide support to teachers who are learning and growing in these ways.

- Professional development can take place in the context of critical colleagueship among teachers.

Add any other significant points that have come up over the course of your work together in this seminar.

Agenda preview Explain that the two activities after the discussion of their homework will give the group a chance to work with these new ideas in the context of other school settings. In the third activity, they will have a chance to think specifically about their own school or district setting.

Facilitator's Notes
OPENING

Module review

This session brings together the aspects of professional development for teachers that participants have been exploring in this module. Listing the major ideas of the module on the board will refresh participants' memories and help to make these ideas more accessible to them during the session.

Besides the four ideas suggested, add any other significant points about professional development that have come up during your work together. For example, you might include the importance of thinking about ongoing opportunities for administrators, as well as teachers, to learn about mathematics learning and teaching.

ACTIVITY 1
Homework Discussion
35 minutes

Materials
Participants' written homework from Session 4
Prepared list of "challenges"

Professional Development in Our Schools

30 minutes | **Sharing personal experiences** One homework assignment that administrators prepared for today was a written memo about the nature of the professional development that teachers in their schools have experienced. Although all are free to keep the specifics of their memos confidential, ask if anyone is willing to share any reflections.

As you prepared your memo and reflected on professional development in your school or district, did you discover anything that surprised you?

If no one volunteers to share, you might ask if anyone is willing to simply describe any professional development programs recently offered. As people share examples, do a quick analysis of these as a group. If the sharing starts to feel random and inconclusive, pick out two or three different examples to compare and contrast.

I hear lots of different types of professional development examples being mentioned. Let's focus for a few minutes on the examples raised by [participant] and [participant]. Let's look for the ways these programs help teachers develop the capacities they need.

THE BIG PICTURE
This discussion helps participants take stock of the professional development activity that has taken place in their schools, focusing not so much on *what activities took place,* but rather on *the nature of the learning that took place.* Researching what has actually happened in their schools reinforces the importance of working with actual data, with an eye to uncovering what might not otherwise have been apparent.

5 minutes | **Setting aside tensions** Call attention to the challenges participants face in their own schools that have been brought up in previous sessions. While acknowledging these, encourage group members to put their concerns on hold as they consider the experiences of other systems.

TRANSITION | Advise participants that they are now going to read about several examples of teacher learning in nontraditional contexts.

Facilitator's Notes
Activity 1

Sharing personal experiences

Participants who feel comfortable sharing might reveal that their professional development activities for teachers have generally been one-time events with a transmission mode of delivery. They may mention workshops intended to convey specific teaching procedures, to familiarize teachers with particular materials, or to introduce a particular mode of record keeping. You might hear that some of the richest learning opportunities came in language arts, where teachers explored a process approach to teaching writing. If this comes up, try to draw parallels with mathematics. When teachers are learning to work with standards-based mathematics instruction, what capacities do they need to develop? How might these capacities be enhanced by a similar experience in a different subject? Participants might also mention professional development that helped them or their teachers look at issues of diversity, focusing on a reorientation toward working with *all* children.

Setting aside tensions

This discussion of the written homework might raise questions in administrators' minds about how to solve newly identified problems. For example, some may be concerned about how to move from a one-size-fits-all professional development approach to one that works well for a range of teachers whose professional development needs are different. Others may be grappling with a newly perceived need to develop a long-term plan for professional development, as opposed to offering a series of discrete experiences. They may be thinking about how to reach all current teachers in the system, including new teachers.

An orientation toward problem solving could make it difficult for administrators to consider the ideas in the first two activities of this session, particularly if your group includes administrators whose schools face immediate critical concerns such as high drop-out rates or low scores on standardized tests, and administrators for whom resources of time and funding are scarce.

You can help participants put their immediate concerns on hold by explaining that later in the session they will have a chance to think specifically about their own school or district settings. Organizing the session this way helps administrators avoid jumping quickly into short-term problem solving and instead gives them some ideas and grounded images of what it means to consider a long-term plan for professional development.

ACTIVITY 2
Readings
60 minutes

Materials
Reading 14
Flip-chart paper and colored markers for each group

Vignettes of Individuals Moving in the Stream

25 minutes	**Small-group discussion of vignettes** Have participants turn to Reading 14, which contains four vignettes about teachers. Ask them to form groups of three. Distribute flip-chart paper and colored markers, and assign each group one of the four vignettes: Valerie and Marta, Marina, the fifth-grade team, or Paul. Explain that while reading the vignettes, participants should think about the experiences and influences that affect the learning trajectories of these individuals.

Give participants about 25 minutes to read and discuss their vignette. Call attention to the directions on page 183 of the readings as you describe the three parts of their assignment.

1. *Read the vignette. Describe the need or interest for teacher growth that is articulated in the vignette and the way that need or interest was initially identified.*

2. *Describe the ways this need or interest was addressed. Record the process on flip-chart paper with words, arrows, and/or drawings.*

3. *Brainstorm other possible ways to support this growth. Record these ideas on the flip-chart paper as well.*

THE BIG PICTURE
This activity helps administrators see that any one teacher's learning takes place over time and is often the result of a number of interactions with colleagues. One great benefit of a culture of critical colleagueship is that ideas begin to flow laterally, in a number of directions and across different boundaries, as teachers have increased opportunities for meaningful shared experiences and interchanges.

35 minutes	**Whole-group sharing and discussion** Invite groups of three to share their vignettes and diagrams with the whole group. Follow this sharing with a discussion of the themes that emerged. Use the following questions to guide the discussion:

What are some ways that teachers get inspired to change their mathematics teaching?

In these vignettes, what allowed these teachers to overcome some of the roadblocks they encountered?

How does all of this relate to the notion of critical colleagueship?

TRANSITION	Tell participants that they will next look at the homework reading about a large-scale professional development project in an urban district that attempted to reach all teachers in the system.

Facilitator's Notes
Activity 2

This activity highlights the relationship between the learning that a teacher needs or wants to do and the various contexts that might provide that opportunity. As administrators begin thinking about the learning that teachers want and need, they may realize that there are many ways to support that learning.

Through these four vignettes, administrators see that teachers' professional lives are full of informal learning possibilities; formal programs are not the only place for professional development. Furthermore, many opportunities for professional growth can be initiated and carried out by teachers themselves. This is an important idea for administrators to consider; otherwise, they may feel overwhelmed by the wide range of professional development needs of their teachers. The value of critical colleagueship, as discussed in the last session, is also highlighted here.

Small-group discussion of vignettes

Circulate to observe small groups as they respond to the three parts of this assignment.

- **Describe the need or interest for teacher growth that is articulated in the vignette and the way that need or interest was initially identified.**

Valerie and Marta A new principal provided the impetus for Valerie and Marta's growth. These two teachers' initial goal was to teach in a more developmental way. Their first move was to provide open-ended problems and allow students to explore and come up with different ways of solving them. As a result, Valerie and Marta discovered a need to gain some experience working with open-ended problems. They also became interested in taking part in scoring student work on assessment tasks, which in turn led them to want to learn more about their own students' understanding of the math.

Marina The impetus for Marina's professional growth was her own perception that in order to work with the full range of students in her class, she needed to deepen her understanding of the mathematics she was teaching. Through her course work, her interest in her students' thinking grew; she also became fascinated by her own development as a learner of mathematics.

The Fifth-Grade Team This group came together initially because of their common interest in exploring the activities in a new curriculum, specifically to learn more about this approach to teaching mathematics. They tried out the activities, first for themselves and then in their classrooms. When they met afterwards to discuss what happened, they discovered a shared interest in how students responded to the activities and the ideas they raised.

Paul A colleague provided the impetus for Paul to begin to examine his practice, urging him to observe a clinical interview. As a result, he discovered that children, when asked certain kinds of probing questions by a teacher, reveal interesting ways of thinking. This led him to wonder how his own students were making sense of mathematics and to consider the on-your-feet decisions teachers must make in response to students' ideas.

- **Describe the ways this need or interest was addressed.** You will likely see a variety of visual representations of the ideas described in each vignette. Look for recognition of the roadblocks encountered and an appreciation of how the learning spread to other aspects of the teachers' practice, to other teachers in the school or district, and to school administrators.

- **Brainstorm other possible ways to support this growth.** As participants consider this matter, remind them of the discussion in the previous session about Readings 5–9, five different approaches to professional development. Some of those models may apply here.

Whole-group sharing and discussion
As groups share their visual representations with the entire class, weave into the discussion some of the following ideas:

- the variety of different reasons these teachers became actively involved in their own professional growth
- the ongoing and evolving nature of the process in each case
- the different professional development interests and needs of the teachers
- the different ways these needs were addressed

Such observations can help administrators envision realistic possibilities for professional development to address a range of different needs in their schools.

◆ **What are some ways that teachers get inspired to change their mathematics teaching?** A new principal inspired Valerie and Marta to change their teaching. Marina's interest grew out of her own dedication and desire to work productively with a wider range of students in her classes. For the fifth-grade team, it was a new curriculum that motivated them. For Paul, it was a colleague he respected who tapped into his existing interest in students' thinking about literature.

◆ **What allowed these teachers to overcome some of the roadblocks?** The teachers in these vignettes encountered such roadblocks as hostile colleagues, child-care difficulties, parent concerns, and a lack of a supply closet. A common theme in overcoming these difficulties was the interest and support of a school administrator—a new principal, a math supervisor, a math coordinator—and the interest and support of colleagues.

◆ **How does this relate to the notion of critical colleagueship?** With the exception of Marina, the teachers in these vignettes had colleagues who fueled their interest in changing their practice and who participated in an ongoing way. A school or district that fosters critical colleagueship can provide fertile ground for learning to take place among all members of the community.

Administrators must recognize that individual or small-group efforts such as the ones described in these vignettes need the recognition and support of the school if they are to become anything more than scattered efforts. Examples of structures that the school can provide include working with a common and carefully selected curriculum; regular grade-level (as well as across-grade-level) meetings, where substantial discussions about learning and teaching take place; supervisory practices that support the direction the school or district wants to be taking; and other forums, both formal and informal, for collegial sharing and discussion.

ACTIVITY 3
Readings
65 minutes

Materials
Reading 10

A System on the Move: Project IMPACT

10 minutes | **Reviewing the reading** Ask participants to turn to Reading 10, "Empowering Children and Teachers in the Elementary Mathematics Classrooms of Urban Schools," which they have read for homework. To begin, have them reconstruct the main points of the article. You might ask,

What were the key elements of the professional development plan for Project IMPACT?

What was involved in making it happen?

This will help refresh everybody's memory about the content of the article and clarify any different understandings about the project itself.

15 minutes | **Benefits and trade-offs** Ask participants to consider the apparent benefits and trade-offs of a plan such as this one, which attempts to reach all teachers with a single program.

40 minutes | **Refining the ideas** Finally, ask participants to suggest possible modifications or adaptations to the Project IMPACT plan, in order to meet the needs of a wider range of teachers. You might say,

This program was in effect in 1990. Given what we know now about teachers' learning, how might we refine the ideas set out in the plan for Project IMPACT? Remember that we still need to incorporate the needs of school or district systems and build on the possibilities that this systemwide work presents, while at the same time we need to be more individually responsive.

In addition to drawing on administrators' own learning experiences within and beyond the *Lenses on Learning* course, remind them to consider the ideas articulated in the other readings they did in preparation for this session (Readings 11–13). Also, there is time in this activity for discussion of ongoing challenges in participants' own situations, if they come up.

Facilitator's Notes
Activity 3

Project IMPACT, described in Reading 10, is a model of professional development for supporting standards-based mathematics that began in 1990 and continued for about six years. Administrators consider how this plan might be adapted to meet the needs of a wide range of teachers, reflecting some of the ideas explored in the previous activity and the first four sessions of this module.

Reviewing the reading

Participants may mention the following elements of the professional development plan for Project IMPACT:

- The project was a collaborative venture between a university and a public school district.
- All teachers of mathematics in the cooperating schools participated.
- All teachers used the same mathematics curriculum.
- The plan included a summer in-service program for all teachers.
- There was an on-site mathematics specialist in each school.
- Teachers had manipulative materials in each classroom.
- One hour was set aside each week for grade-level mathematics meetings, which were used for both lesson planning and instructional problem solving.

As to what was involved in making it happen, participants may mention that the understanding and support of school administrators are critical to the success of such a program. Alignment between what teachers are trying to do and what administrators believe teachers should be trying to do is essential; when teachers are changing their practice, administrators must focus on different things as they conduct classroom observations or choose a new curriculum.

Although many elements of the program are important to its success, according to the author of this article, "what is critical is some mechanism to support change, to foster implementation, to promote reflection, to applaud efforts, and to challenge further growth. In Project IMPACT, that mechanism happens to be the [math] specialist."

Benefits and trade-offs

This program attempts to reach all teachers with a single program. Following are some possible benefits that participants may note: A systemwide approach might move all but the most reluctant teachers to begin making changes, because a new set of expectations establishes a new norm of professional involvement. Teachers might feel much less "out on a limb" if the changes they were making were also being implemented school- or districtwide. Teachers would have more in common with one another, because of their shared experiences in the professional development offerings and in their efforts to change their teaching accordingly. As a result, team teaching and classroom observations would be easier. Changing the classroom culture could be easier since students would be experiencing the same changes from year to year with all their teachers.

Administrators might mention a number of trade-offs as well. They might bring up the difficulty of establishing good "fits" between standard professional development offerings and the range of teacher experience, strengths, and interests in their schools. They might be concerned about teacher resistance to a mandated system. They might worry that the spark of individual initiative could be quelled by a broad-based effort such as this one. They might wonder how to develop and maintain a quality experience in so large a program. If someone does not mention it, remind participants of "contradiction 2" in Lord's article on critical colleagueship, which states that an effort to reach all teachers in a district in order to achieve widespread results might result in "a precariously thin staff development program and little real change."

Refining the ideas

The question about refining the plan for Project IMPACT invites administrators to draw on their own learning and experience, within and beyond this course, as they share their current understanding of professional development. This is a good time for you, as facilitator, to see which new ideas administrators are incorporating into their thinking.

One idea relevant to this discussion, mentioned above, is that it can be difficult to establish a good fit between individual teachers' needs and a single professional development program designed to reach all teachers in a district.

Encourage participants to think about Readings 11–13; ideas in these articles may suggest some refinements. Following are some refinements you may hear mentioned:

- **Making plans for changing support needs**
Administrators whose districts have already chosen a curriculum may want to think about the range of pre-implementation, early implementation, and ongoing implementation needs. Those whose districts have not yet settled on a curriculum may still find this construct helpful in thinking about the support needs of individual teachers.

- **Developing a cadre of teacher leaders**
Administrators might find themselves thinking about teacher leadership for mathematics education in new ways. What roles can teacher leaders play in planning for professional development for mathematics in the schools? What roles might they play in supporting the learning that takes place in grade-level meetings, in cross-grade-level meetings, or for individual teachers in the classroom? How will these leaders be identified? How will they be compensated for their time and commitment? How will they be supported?

Teacher leaders will need ongoing support. They need opportunities to continue to work on their own teaching practice, to become familiar with the mathematics being taught at all other grade levels, and to develop their capacity to work the range of teachers, particularly by exploring the processes that support adult learning. These opportunities would be integral components of a long-term professional development plan.

- **Finding ways to support the learning of other administrators in the school and district**
Participants might reflect on the important components of their own learning, how these have affected their work as administrators, and how other administrators in their school or district might be encouraged to undertake similar learning experiences.

- **Making allowances for new perspectives on time**
Administrators may find themselves thinking differently about how long it is likely to take teachers to understand and make the changes that are central to teaching, learning, and assessing mathematics in concert with NCTM's *Principles and Standards for School Mathematics*. They may also consider ways to shift the balance and configuration of instructional time and planning time in their schools in order to better support the professional development of teachers.

Other refinements that administrators might suggest may relate to aspects of the *Lenses on Learning* course that have particularly stretched their own thinking—clinical interviews, mathematics investigations, particular video clips or readings, discussion questions—and that might be shared with other members of their school community.

Practical considerations

Earlier in this module, you have probably heard participants mention the specific issues and challenges they face in offering new kinds of professional development for teachers in their school districts. It was suggested that you keep lists of their concerns for later discussion. There is sufficient time built into this activity to accommodate such a discussion if you feel it would benefit participants in your course.

Administrators may want to talk about their own situations and to share actual strategies, discuss logistics, and trade concrete suggestions. It is important for you to acknowledge the very real, complex situations they face. At the same time, you can encourage them to explore how the ideas they have been discussing in this course—about teacher learning and the variety of ways to support that learning—can help them view their problems in a new way and find paths to solutions that they may not have seen before. That is, help them ground their discussions of the practical in the conceptual base that they have been exploring throughout this module.

CLOSING
15 minutes

Bridging to Practice

Invite participants to respond to the Bridging to Practice questions.

1. Pick an idea that came up today that you found particularly interesting. What is your current thinking about this idea?

2. Where is your school now with regard to this idea?

3. What are one or two things that you will go back and pursue, to move yourself and/or your school along with this idea?

Because this is the final session in the module, participants may want to reflect more globally on professional development in their schools from their perspective as instructional leaders. The following two questions can guide them in that writing:

4. What things continue to puzzle you about meeting the learning needs of teachers in your schools through professional development?

5. What new possibilities do you think you might be interested in exploring, even if there are significant barriers to implementation?

Resource List

Journal articles and chapters in books

Burns, Marilyn. 1998. Who needs tips tables? In *Facing an American phobia.* Sausalito, CA: Math Solutions Publications.

    Burns appeals to the reader's own experience of doing math in real-life scenarios that involve estimation and number sense and stresses that students in the classroom need to be asked to think and reason at all times, not merely to learn rules and practice them.

Grant, Catherine Miles. 1999. Beyond tool training: Professional development for elementary mathematics teachers. *Learning and Leading with Technology* 27 (3): 42–47.

    Grant proposes that through collaboration with their peers and with the support of a comprehensive schoolwide approach to professional development, elementary math teachers can deepen their understanding of mathematics, teaching, and learning while enhancing their information technology skills.

Russell, Susan Jo. 1998. Mathematics curriculum implementation: Not a beginning, not an end. *Hands On!* 21 (1): 6–9.

    Russell discusses the role curriculum can play in systemic reform. She focuses on the elementary mathematics curriculum for grades K–5 and examines the experiences of several school systems as they implement new curriculum materials. She also highlights the roles of professional development and parents in the implementation process.

Schifter, Deborah, and Catherine Twomey Fosnot. 1993. The rug pulled out from under her: Sherry Sajdak. In *Reconstructing mathematics education: Stories of teachers meeting the challenge of reform.* New York: Teachers College Press.

    This case study profiles one teacher and the struggles and emotional turmoil she endured as she worked to integrate new pedagogical and mathematical ideas into her mathematics teaching. The story makes clear that for some teachers, this process can raise many conflicts and may entail a clash of principles and practices. In this case, the authors describe how one teacher came to terms with these conflicts and took control of her own teaching practice and professional growth.

Signet, Mary. 1996. Food for thought and talk. In *What's happening in math class? Volume 1: Envisioning new practices through teacher narratives,* edited by D. Schifter. New York: Teachers College Press.

    This essay chronicles one teacher's attempts to engage four third-grade students who were performing below their grade level in mathematics. Signet describes how she was able to start children talking about mathematics by first getting them to talk about something else that interested them. The paper illustrates the strategic ways in which teachers can build on what children already know and on the interests children bring to the classroom in order to make connections to more substantive mathematical ideas.

Stigler, James, and James Hiebert. 1997. Understanding and improving classroom mathematics instruction: An overview of the TIMSS video study. *Phi Delta Kappan* 79 (1): 14–21.

The Third International Mathematics and Science Study (TIMSS) developed a detailed analysis and comparison of eighth-grade mathematics teaching in three countries: Germany, Japan, and the United States. In this brief overview of the TIMSS use of videotape to study teachers at work, the authors discuss implications of that video study for the improvement of classroom mathematics teaching in the United States.

Toney, Nora L. 1996. Facing racism in mathematics education. In *What's happening in math class? Volume 2: Reconstructing professional identities,* edited by D. Schifter. New York: Teachers College Press.

The author of this essay, an African-American teacher, chronicles her development as mathematical thinker despite her continual confrontations with both explicit and tacit forms of racism in both her own education and her work as a teacher. She draws from her own experiences to offer guidelines and principles for teaching minority students as well as understanding and respecting their different learning styles and needs.

Books

Eisenhower National Clearinghouse. 1998. *Ideas that work: Mathematics professional development.* ENC Online. <www.enc.org/professional/learn/ideas/>

This booklet offers an overview of professional development in mathematics, useful for professional developers, teachers, evaluators, and funders. It outlines fifteen specific strategies for professional development, describes seven successful programs, and provides a resource list of technical assistance providers and additional programs.

Fennema, Elizabeth, and Barbara Scott Nelson, eds. 1997. *Mathematics teachers in transition.* Mahwah, NJ: Lawrence Erlbaum.

The essays in this book address the need for professional development leaders and policymakers to have scholarly knowledge that influences teachers to bring mathematical instruction into alignment with standards-based mathematics reform efforts. The book presents theoretical perspectives for studying, analyzing, and understanding teacher change; descriptions of contextual variables to consider when studying change; and descriptions of professional development programs that resulted in teacher change.

Heaton, Ruth M. 2000. *Teaching mathematics to the new standards: Relearning the dance.* New York: Teachers College Press.

Ruth Heaton, an elementary teacher with ten years of experience, describes and traces her efforts to change the way she teaches in response to current national reforms in mathematics education. In chapters that alternate the teacher's-eye view with that of the researcher, Heaton interweaves explorations of theory with its practical application as demonstrated in actual classroom vignettes. (This is a more in-depth account of the process reported in Reading 2, "Learning to Hear Voices," by Heaton and Lampert.)

Nelson, Barbara Scott, ed. 1995. *Inquiry and the development of teaching: Issues in the transformation of mathematics teaching.* Newton, MA: Center for the Development of Teaching, Education Development Center.

The five essays in this collection report on projects involving work with teachers to help them examine their fundamental beliefs about learning, teaching, and the nature of mathematics, to deepen their mathematics knowledge, and to reconstruct their teaching practice from within a new conceptual framework. The papers provide both conceptual and practical explorations of dimensions of teacher change, including mathematics knowledge, the role of affect in the process of teacher change, the role of curricular materials, and the effects of a new inquiry-based teaching culture.

Schifter, Deborah, ed. 1996. *What's happening in math class? Volume 1: Envisioning new practices through teacher narratives.* New York: Teachers College Press.

The essays in this collection, written by teachers working with constructivist principles and new pedagogical methods to transform their teaching according to the NCTM *Standards*, provides insightful and powerful narratives that illustrate what is involved when teachers work to enact reform principles in their practice. It includes five analytic essays written by teacher educators that highlight some of the broader issues raised in the teachers' essays.

Schifter, Deborah, ed. 1996. *What's happening in math class? Volume 2: Reconstructing professional identities.* New York: Teachers College Press.

In this collection of essays, teachers explore their struggle to understand constructivism and its application to learning mathematics. The narratives are complemented by four essays, written by teacher educators, that examine some of the challenges posed by the new mathematics pedagogy for teachers' professional identities.

Schifter, Deborah, and Catherine Twomey Fosnot. 1993. *Reconstructing mathematics education: Stories of teachers meeting the challenge of reform.* New York: Teachers College Press.

Through case studies, Schifter and Fosnot explore the struggles, doubts, and successes of teachers who are working to bring a constructivist understanding of mathematics learning to their day-to-day practice. In each case, the authors consider the teachers' intentions in designing a particular mathematics activity, the instructional decisions she makes as the children engage in it, and her reflections afterward.

Videotapes

Mathematical inquiry through video: Tools for professional growth. 1997. A package of videos and teacher professional development materials developed by BBNT Solutions LLC.

This video package presents ten cases of middle-school mathematics classrooms. The videos depict teachers who are working to change their teaching practice according to the NCTM *Standards* and highlight the challenges, ideas, and issues these and other teachers face in this process. Each video is accompanied by a facilitator's guide that includes a description of the video; background information on the school, teacher, and classroom shown in the video; suggested workshop features with video-related mathematical and pedagogical issues and activities; and a complete transcript of the video. For more information on this series, contact Fadia Harik at harik@attbi.com or fadia.harik@umb.edu.

Mathematics: Assessing understanding. 1993. A series of videotapes for staff development. Created by Marilyn Burns. White Plains, NY: Cuisenaire.

This series, with three videotapes and an accompanying teacher's discussion guide, shows a collection of individual assessments of mathematical understanding with students ages 7 through 12. All of the assessments address students' ability to estimate, reason numerically, and compute in problem-solving situations. The one-on-one interviews model for teachers the kinds of questions that are useful for gaining insights into how students are thinking and what they understand. Mathematical topics included are number sense and the place-value structure of our number system, estimation, numerical reasoning, and computation with whole numbers and fractions.

Relearning to teach arithmetic. 1999. A professional development package developed at TERC by Susan Jo Russell, David A. Smith, Judy Storeygard, and Megan Murray. Four videotapes, two study guides. White Plains, NY: Dale

Seymour Publications, an imprint of Pearson Learning Group, a division of Pearson Education, Inc.

This two-part video series provides teachers a structured opportunity to explore the ways children develop facility with the four operations (addition, subtraction, multiplication, and division) and the ways teachers can foster the development of this facility. The first package focuses on addition and subtraction; the second, on multiplication and division. Each package contains two videos and a study guide that outlines six professional development sessions. During each session, teachers view and discuss segments of the tapes and work on related mathematics problems. The series is adaptable to a variety of settings, such as after-school sessions, release-day professional development sessions, or summer staff-development experiences. The materials may also be integrated into a longer course or seminar on the teaching and learning of elementary mathematics.

Talking mathematics. 1996. A professional development resource package developed by TERC, Cambridge, MA. Portsmouth, NH: Heinemann.

This package includes a videotape program, a resource guide for staff developers and university instructors, and a book for teachers who are interested in supporting talk and mathematical inquiry in their classrooms. The goal of the package is to provide teachers and staff developers with resources that can help them cultivate good mathematical discourse. The video program specifically consists of an introductory videotape, four 20-minute videotapes on aspects of children's talk, six short classroom episodes, and a 20-minute summary of a *Talking Mathematics* teacher seminar. The package can be adapted to a variety of professional development settings.

Teaching math video libraries. 1995. A series of videotapes for staff development, produced by WGBH. South Burlington, VT: The Annenberg/CPB Math and Science Collection.

This video series provides visual examples of standards-based teaching and learning. Four libraries are available: K–4, 5–8, 9–12, and a K–12 assessment library. The grade-level libraries each include a set of content-standard videos, a set of process-standard videos, and guidebooks. The assessment library provides case study videos that examine assessment issues in two different classes and a sequence of vignettes that show a variety of assessment techniques from several classes. This collection of video libraries provides grounded images of what classrooms may look like when teachers are developing their teaching in accordance with the NCTM *Standards*.